BASIC CIRCUIT ANALYSIS FOR ELECTRONICS
Using Electronics Workbench®

Lorne MacDonald

The Technical Education Press
Chico, California

BASIC CIRCUIT ANALYSIS FOR ELECTRONICS
Using Electronics Workbench®

Copyright © 1998 by The Technical Education Press. All rights reserved. Printed in the United States of America. No part of this publication may be reproduced, stored in retrieval systems, or transmitted, in any form or by any means, electronic, mechanical, photocopying, recording, or otherwise, without the prior written permission of the publisher. The Technical Education Press, P.O. Box 397, Chico, CA 95927.

ISBN: 0-911908-26-9

Manufactured in the United States of America

Preface

This book is a supplement to **Basic Circuit Analysis for Electronics Through Experimentation 3/e**, and the Electronics Workbench® program, which makes available a virtual electronics laboratory complete with components and test equipment, is used to work the experiments. So, while the basic book is dedicated to the construction and testing of circuits in a real hardware laboratory, **Basic Circuit Analysis for Electronics Using Electronics Workbench®** is dedicated to the construction and testing of circuits on a computer. Therefore, the concepts mastered in hardware laboratory can be mastered on the computer laboratory without going near an actual laboratory.

Too, while the hardware laboratory is necessary to provide "hands on" experience needed in the mastery of electronics, the EWB program allows students to concentrate on concepts rather than on circuit construction. This speeds up the learning process making it possible to do all of the laboratory exercises, both ways, in the time formerly needed to work only the hardware laboratory exercises. Also, to help speed the process, all of the circuits encountered in the computer laboratory exercises are available on floppy disks.

An advantage of using the computer laboratory is that the component values are exact, so the calculated and measured values will be exact. Any discrepancies that might occur, such as an improperly connected circuit or an incorrect component value, is immediately detected. Also, of advantage is the availability of a current source and a Bode plotter, neither which is readily available in a real laboratory. Thus, the constant current source can be used to help clarify the concept of current splitting, and the Bode plotter provides an immediate frequency response without having to use point-by-point plotting. In all respects, all the concepts that can be learned in a real laboratory can be learned in the virtual laboratory on the computer.

The availability of computer laboratory exercises is particularly useful to students working at home or in a distance learning situation. They can obtain the same block of knowledge needed for the more advanced electronic course work by using this computer laboratory based program of instruction.

Acknowledgments

A debt of gratitude is owed to the many former students, who class-tested the experiments, and to colleagues, who provided encouragement and valuable suggestions. I am especially grateful to Leon Lucchesi of Truckee Meadows Community College, George Bramlett of College of San Mateo, and Roman Bysh and Joe Koenig of Interactive Image Technologies LTD. I would also like to acknowledge LeRoy Olson for his editorial advice and assistance. Lastly, I wish to thank my wife Annette, who read the manuscript and provided encouragement.

Lorne MacDonald
1998

Electronic Circuit Analysis Series
from
The Technical Education Press

BASIC CIRCUIT ANALYSIS FOR ELECTRONICS
Through Experimentation—3rd Edition
    by Lorne MacDonald

BASIC CIRCUIT ANALYSIS FOR ELECTRONICS
Using Electronics Workbench®
    by Lorne MacDonald

BASIC SOLID STATE ELECTRONIC CIRCUIT ANALYSIS
Through Experimentation—3rd Edition
    by Lorne MacDonald

PRACTICAL CIRCUIT ANALYSIS OF AMPLIFIER
    by Lorne MacDonald

PRACTICAL ANALYSIS OF ADVANCED ELECTRONIC CIRCUITS
Through Experimentation—2nd Edition
    by Lorne MacDonald

DIGITAL CIRCUIT LOGIC AND DESIGN
Through Experimentation
    by Darrell D. Rose

Table of Contents

1. Introduction to Electronics Workbench.. 7
2. Resistors and Resistance Measurements... 11
3. Ohm's Law.. 20
4. Series Circuits.. 26
5. Parallel Circuits.. 32
6. Series-Parallel Circuits.. 37
7. Series-Aiding Two Power Supply Circuits.. 44
8. Voltage Variable Circuits and Applications.. 52
9. Thevenin's and Norton's Equivalent Circuits and Maximum Power Transfer... 59
10. Complex Circuit Analysis... 66
11. Circuit Loading Effects.. 73
12. Capacitors in DC Circuits.. 78
13. Introduction to AC Sine Wave Analysis and Oscilloscope Measurements........ 87
14. Combined DC and AC Circuit Analysis... 97
15. Capacitors in AC Sine Wave Circuits... 104
16. Inductors in AC Sine Wave Circuits... 112
17. Transformers... 119
18. Low Pass Filters.. 127
19. High Pass Filters... 133
20. Bandpass RC Filters.. 139
21. Bandpass and Band Reject LC Filters... 146
22. Complex Waveshapes.. 153
23. Square Wave Testing of the Corner Frequencies...................................... 157

Dedicated to My Daughters
Lynn
Leslie
Lisa
Louise
Laura

CHAPTER 1
INTRODUCTION TO ELECTRONICS WORKBENCH®

INTRODUCTION

Electronics Workbench® is a program that creates a virtual electronic laboratory on the computer. It is complete with components and test equipment. So, all of the circuits in **Basic Circuit Analysis For Electronics Using Electronics Workbench®** can be constructed and tested and the concepts mastered without actually going near a real hardware laboratory. While the hardware laboratory is necessary to provide "hands on" experience needed in the mastery of electronics, the EWB program allows the student to concentrate on concepts rather than on circuit construction. If possible, however, the laboratory exercises should be done in both the hardware laboratory and on the computer

STARTING THE EWB PROGRAM

With the computer turned on, and the Electronics Workbench® program installed on the hard disk, select programs and Electronics Workbench®. The main window will appear as shown in Figure 1-1. The top row of commands on the main window is the Menus bar containing the program commands. The second row contains the Circuit toolbar icons for the easy access to the most widely used program commands. The third row is the Parts Bin toolbar containing all the components needed to construct the circuits, measuring instruments that allow the constructed circuits to be measured, and a power switch to activate the circuits.

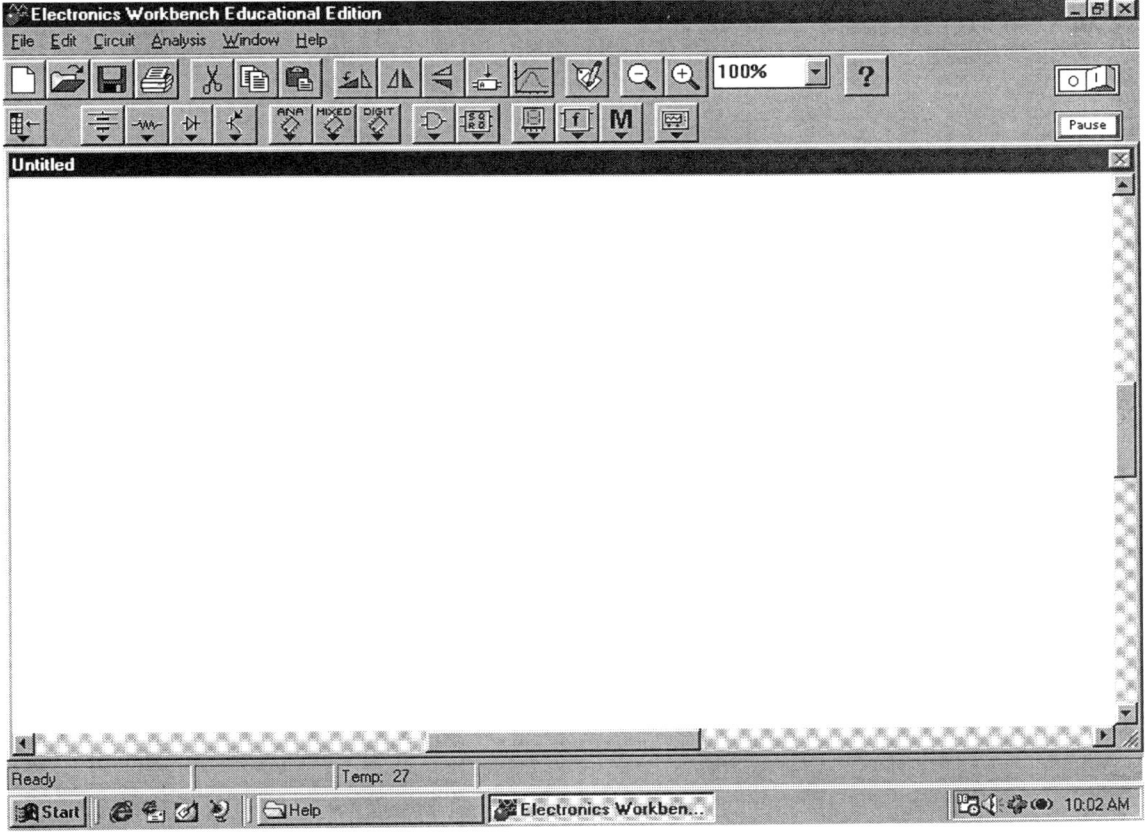

FIGURE 1-1: Main window of Electronics Workbench®

MENUS BAR

The Menus bar for the EWB program contains File, Edit, Circuit, Analysis, Windows, and Help menus. Clicking on file in the Menus bar provides the File menu shown in Figure 1-2. Clicking on Edit, Circuit, Analysis, Window, or Help provides each of these remaining menus.

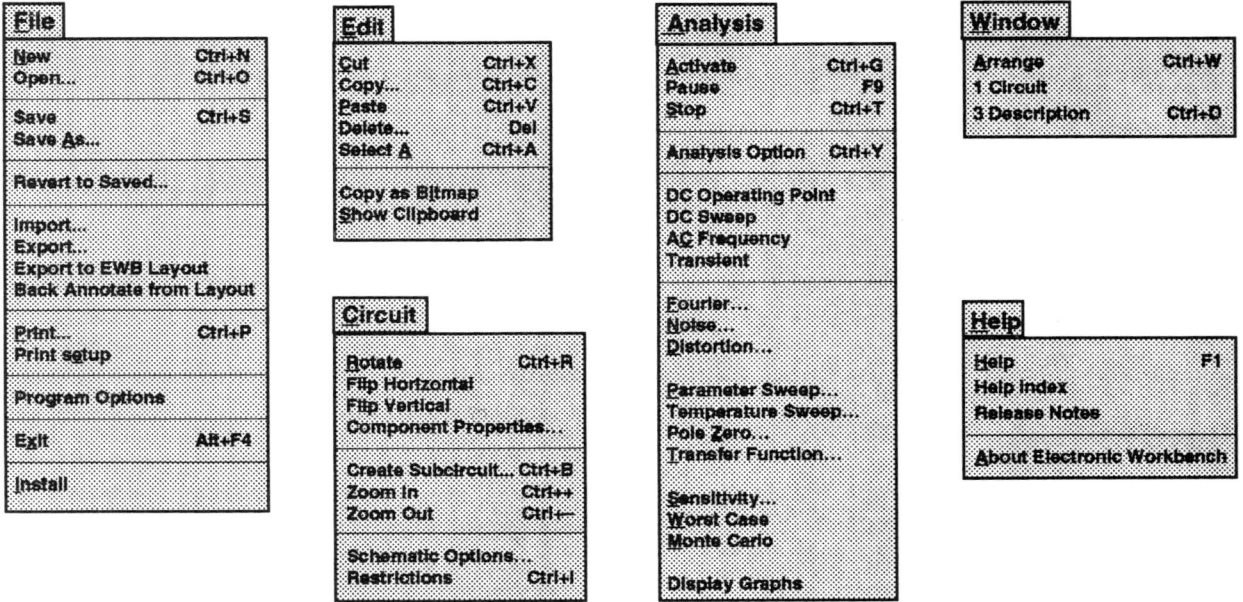

FIGURE 1-2: Menus bar for EWB program.

CIRCUIT TOOLBAR

The most widely used menu commands are available using the Circuit toolbar icons located in the second row of commands. These commands include New, Open, Save, Print, Cut, Copy, Paste, Rotate, Flip Horizontal, Flip Vertical, Create Subcircuit, Display Graphs, Component Properties, Zoom Out, and Zoom In. A captioned representation of the Circuit toolbar is shown in Figure 1-3.

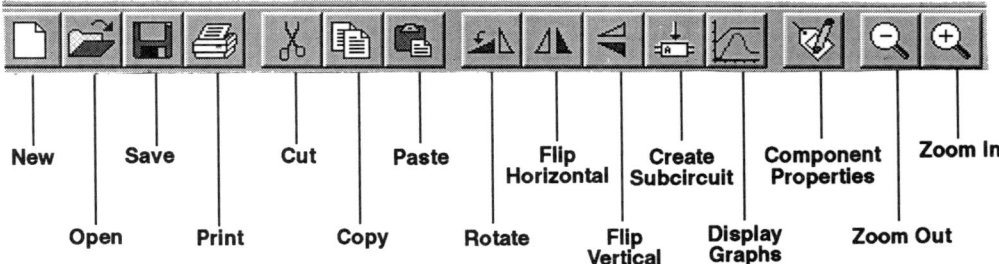

FIGURE 1-3: The Circuit toolbar.

A brief description of each icon follows:

New: Opens a clear workspace, or clears the workspace, prompting you to save the existing file.

Open: Allows a previously created circuit file to be retrieved. Clicking on open will provide a dialog box that identifies the drive/directory or disk/folder where the file is located.

Save: As the word implies, saves the circuit in the workspace by writing it to the disk.

Print: Prints the circuit on the workspace. Print opens a dialog box that allows options such as the schematic and test equipment to be printed. Analysis options can also be printed.

Cut: Removes a component, or a portion, or all of the circuit from the workspace. However it has no effect on the instruments. All objects cut from the workspace are placed on the clipboard, so they can be pasted somewhere else in the circuit. However, placing the component on the clipboard eliminates what was previously pasted on the clipboard.

Copy: Allows the circuit file to be duplicated and removed from the workspace. Again it is saved on the clipboard for easy pasting on the workspace.

Paste: The command pastes the copied or cut contents of the clipboard onto the workspace.

Rotate: As the word implies, rotates the components 90° at a time. For example, a horizontal resistor can be rotated to a vertical position using the rotate command.

Flip Horizontal: As implied, flips selected circuit horizontally.

Flip Vertical: As implied, flips selected circuit vertically.

Create Subcircuit: Saves a circuit or a selected part of a circuit as a subcircuit. Subcircuits are typically stored in the Favorites icon to be used at a later time.

Display Graphs: Used to view, adjust, and save graphs and charts.

Component Properties: Used to label and assign values to the selected component.

Zoom Out: Reduces the size of components, circuits, and instruments.

Zoom In: Enlarges the size of components, circuits, and instruments.

PARTS BIN TOOLBAR

The third command row is the Parts Bin toolbar which contains the necessary components and instruments to construct and test a circuit. The icons represent a Sources bin, Basic bin, Diodes bin, Transistors bin, Analog ICs bin, Mixed ICs bin, Digital ICs bin, Logic Gates Bin, Digital bin, Indicators bin, Controls bin, Miscellaneous bin, and the Instruments bin. The contents of each bin is displayed by clicking on their individual icons, which shows the schematic or pictorial representations of the various components and instruments available. A captioned representation of the Parts Bin toolbar is shown in Figure 1-4.

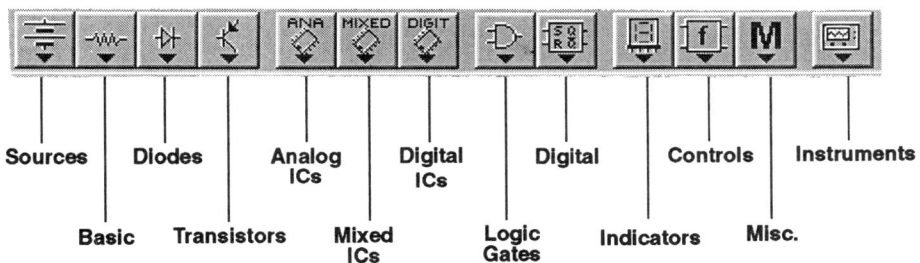

FIGURE 1-4: The Parts Bin toolbar.

In this book we are concerned only with the components and instruments used in the analysis of direct and alternating current circuits. Those components and instruments are found mainly in four of the Parts Bin toolbar bins: the Sources bin, the Basic bin, the Instruments bin, and the Indicators bin. The parts and instruments included in these four bins are shown in Figure 1-5.

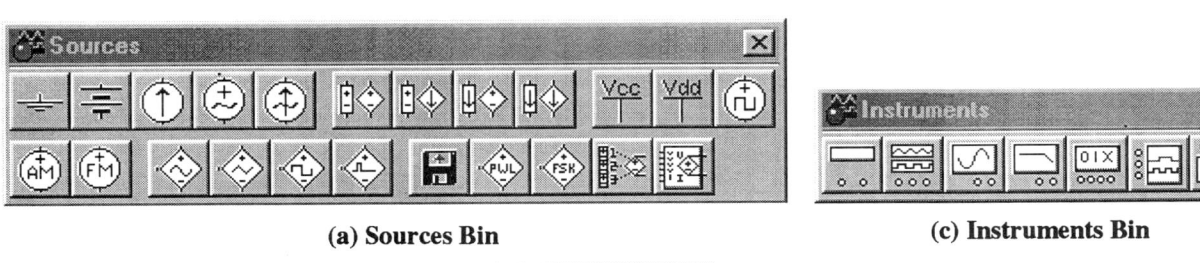

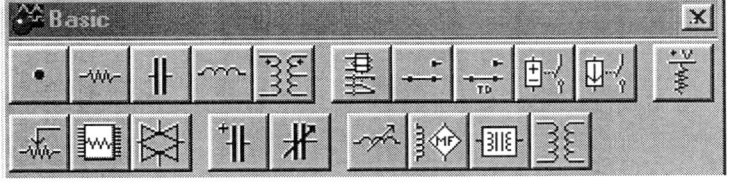

FIGURE 1-5: Four most widely used Parts Bin toolbar bins.

Also, in each of the parts bins we are concerned with only a few of the components. From the Sources bin the ground connection, the battery, the constant current source, and the alternating current source are used. In the Basic bin the connection point, resistors, capacitors, inductors, and transformers, in various configurations, are used. From the Indicators bin, voltmeters and ammeters are used. Finally, from the Instruments bin, the multimeter, function generator, oscilloscope, and Bode plotter are used.

CONSTRUCTING A BASIC CIRCUIT IN THE WORKSPACE

To construct the circuit of Figure 1-6(c), click on the Sources bin icon. When the dialog box opens, drag the battery and the ground to the workspace from the Sources bin. Place each at the desired position, and release the button. Then, click on the Basic bin icon and, when the dialog box opens, drag the two resistors onto the workspace. To rotate the second resistor 90° use the Ctrl + R command from the keyboard. To finish connecting the circuit, connect a wire between one end of the first resistor and the positive terminal of the battery by using the mouse to point, drag, and connect. Then, connect the wires between the resistors and between the second resistor and ground, as shown Figure 1-6(b). To change the color of the wire, double click on the wire, select the wire color, and click OK.

To change the component values to those shown in Figure 1-6(c), change the 12 V battery to 18 V by double clicking on the dc battery. When the dialog box appears change the value to 18 V and click on OK. To label the voltage, V1 for instance, click on label in the dialog box, enter V1, and click OK. To change the resistor values from the default 1 kΩ resistance to 3 kΩ, double click on the resistor. When the dialog box appears set the resistance value to 3 kΩ. To label the R1 resistor, click on label in the dialog box, enter R1, and click OK. Repeat the procedure for the rotated second resistor and set the value to 6 kΩ and label the resistor R2. Extra components or wire that have to be removed are removed by highlighting the component and selecting cut or the Ctrl + X command from the keyboard.

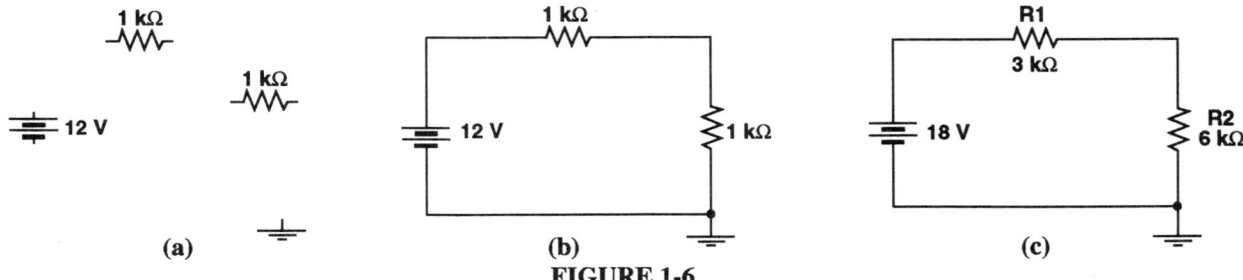

FIGURE 1-6

MORE BASIC INFORMATION

Other basic information necessary to use the EWB program include:
1. To make measurements: Move the instrument onto the workspace from the Instruments bin and connect it to the circuit.
2. To activate the circuit: Click on the power switch located in the upper right hand corner of the window.
3. To move a component around in the workspace: Click on the component with the mouse so it is highlighted. Then, drag the component to the desired location.
4. To remove a component from the workspace: Highlight the component by clicking on it with the mouse. Once the component is highlighted, use the cut command.
5. To remove an instrument from the workspace: Highlight the instrument with the mouse and use the delete key on the keyboard or delete in the Edit menu.
6. Factor is located to the right of the Zoom Out-Zoom In Icons and indicates the percentage of reduction or enlargement.
7. Pause is used to stop the circuit action without turning off the circuit. Clicking on it after the pause resumes the circuit action.

More details can always be obtained by referring to the *User's Guide*, the *Technical Reference* book, or the on-line Help menu. However, learning by doing will make you an expert in a very short time.

CHAPTER 2
RESISTORS AND RESISTANCE MEASUREMENTS

INTRODUCTION

Resistors are the most commonly used component in electronic circuits. The circuit boards in any system such as a television, radio, computer, or radar contain dozens of them. The term resistor comes from the function of the resistors ability to resist the flow of current in a circuit, where the higher the resistance value, the lower the current flow through the resistors. Resistors can be either fixed or variable and fixed resistors can be general purpose or precision. The resistive value of most fixed resistors is identified by color coding, and the most commonly used resistors are the four band general purpose resistors.

RESISTOR SYMBOL

The resistor symbol used in electronics and in the Electronics Workbench® is shown in Figure 2-1(a). To obtain a resistor use the mouse to drag the resistor from the Basic bin to the workspace. Essentially, move the arrow to the Basic bin, click on, find the resistor, and drag the resistor onto the workspace. Then, set the resistor at the desired position and release the button. The default resistance of the resistor is 1 kΩ. To change the resistance value, double click on the resistor symbol. When the component dialog box appears, change the value. For example, to change the value from 1 kΩ to 3.3 kΩ select the value, type in 3.3 kΩ, and click on OK. See Figure 2-1(b). To provide a label for the resistor, say R1, select label, type in R1, and click on OK. See Figure 2-1(c). The labeled resistor is shown in Figure 2-1(d).

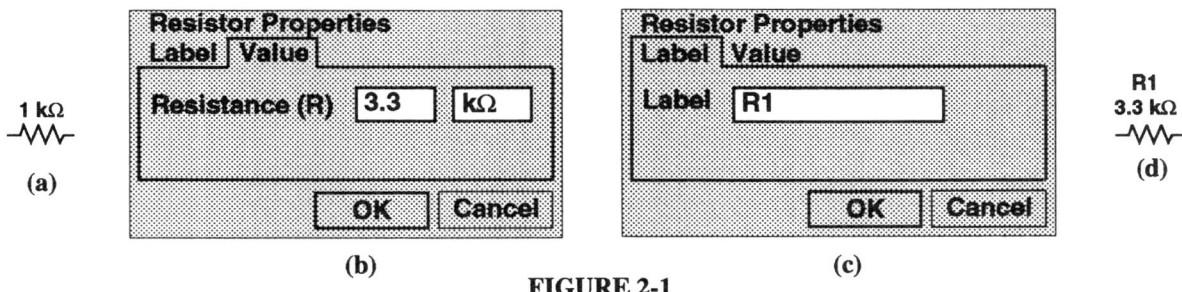

FIGURE 2-1

MEASURING RESISTORS USING ELECTRONIC WORKBENCH®

The resistance of resistors is measured in ohms using an ohmmeter. The instrument mostly used to measure ohms is the multimeter, which has the capability of measuring ohms, volts, and amperes. To obtain the EWB multimeter to measure ohms, move the arrow to the Instruments icon on the Parts Bin toolbar menu, click on, point to the multimeter icon on the instrument shelf, and drag it on to the workspace. Then, double click on the icon to enlarge the meter and, if necessary, switch to ohms (Ω) and dc (—), as shown in Figure 2-2(a). Then point to the terminals on the meter and drag out the wires and attach them across the resistor under test as shown in Figure 2-2(b). To measure the resistance of the resistor use the mouse to click the on-off power switch on the upper right corner of the display, select activate from the analysis menu, or use Ctrl + G on the keyboard.

THE LABORATORY EXERCISE

The laboratory exercise is divided into five sections and each section contains pre-laboratory calculations followed by measurements. The calculations should be done before the actual laboratory exercise so the time on the computer can be used for measurement purposes. If the calculations and measurements are not close, the circuit values or the measurements are wrong or the circuit is not properly connected.

12 — Basic Circuit Analysis For Electronics Using Electronics Workbench

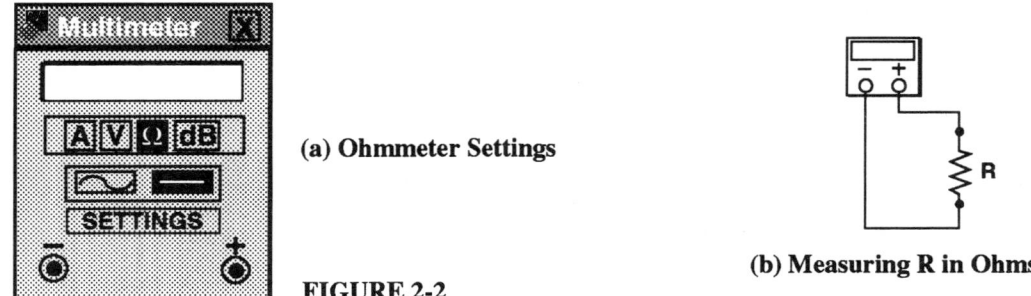

(a) Ohmmeter Settings

(b) Measuring R in Ohms

FIGURE 2-2

In Section I of the laboratory exercise, the color codes in Table 2-2 are used to solve the resistance value and percent tolerance of each resistor and their maximum and minimum values. The resistors are then measured and for four of the eight measured resistors the actual percent tolerances are solved and verified.

In Section II, two resistors in series and three resistors in series are calculated and measured, where $R_T = R1 + R2...R_n$ as shown in Figure 2-3.

(a) $R_T = R1 + R2$

(b) $R_T = R1 + R2 + R3$

FIGURE 2-3

In Section III, two resistor in parallel and three resistors in parallel are calculated and measured, where $R_T = 1/(1/R1 + 1/R2...1/R_n)$, as shown in Figure 2-4.

(a) $R_T = 1/(1/R1 + 1/R2)$

(b) $R_T = 1/(1/R1 + 1/R2 + 1/R3)$

FIGURE 2-4

In Section IV, three and four resistor series-parallel resistors connections are calculated and measured, as shown in Figure 2-5.

(a) $R_T = R1 + (R2 \| R3)$

(b) $R_T = R1 + (R3 \| R4) + R2$

FIGURE 2-5

In Section V, variable resistors are calculated and measured at 90% and 10% of full clockwise rotation, and at mid-rotation as shown in Figure 2-6

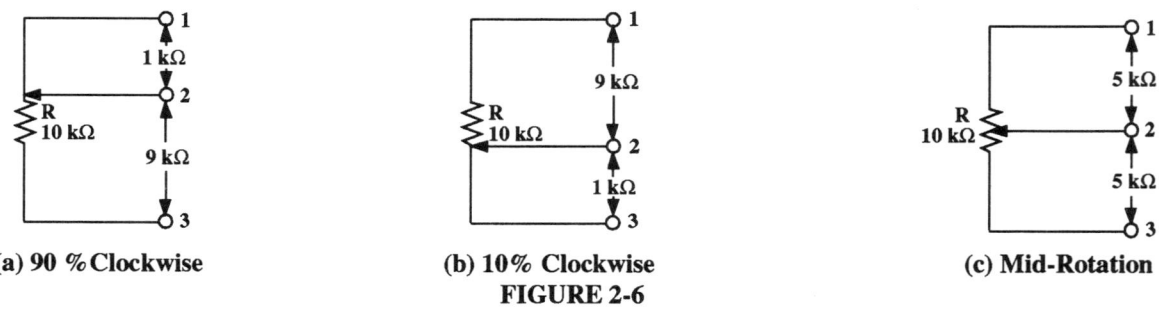

(a) 90 % Clockwise

(b) 10% Clockwise

(c) Mid-Rotation

FIGURE 2-6

LABORATORY EXERCISE

READING ASSIGNMENT: Basic Circuit Analysis for Electronics, 18-31.

EXERCISE OBJECTIVES

To become familiar with:

- The resistor color code.
- The ohmmeter to measure resistance values.
- The equivalent resistance of resistors connected in series, parallel, and series-parallel.
- Variable resistors.

PROCEDURE

SECTION I (Guided Laboratory Exercise): Resistor Color Code

PART A: Pre-laboratory Identification and Calculations

1. From the assortment of resistors listed in Table 2-1, select each successive resistor by color code. Then, based on the color code noted in Table 2-2, determine the resistance value. From the percent tolerance (noted in Table 2-2 and shown in Figure 2-8), calculate the expected upper and lower resistance limit of each resistor. As an example, the resistance values of the BR-BK-BK-Gold resistor have been filled into the first line of Table 2-1. The expected upper limit (maximum ohms) of the $10\,\Omega \pm 5\%$ resistor is $10.5\,\Omega$ and the expected lower limit (minimum ohms) is $9.5\,\Omega$.

2. Resistance and percent tolerance:
   a. Use the color code to determine the value of the next resistor listed in Table 2-1. Record this value into the nominal (coded resistance) ohms column of Table 2-1.

TABLE 2-1 RESISTIORS	IDENTIFY		CALCULATE		MEASURE	VERIFY
	Nominal Ohms	Tolerance Percent	Maximum Ohms	Minimum Ohms	Multimeter Ohms	Percent Tolerance
BR-BK-BK	10 Ω	5%	10.5 Ω	9.5 Ω		
R-R-BR						
BR-BK-R						
O-O-R						
Y-V-R						
BR-BK-O						
BR-G-O						
BR-BK-Y						
BR-BK-G						

14 — Basic Circuit Analysis For Electronics Using Electronics Workbench®

    b. Use the tolerance band on the resistor (indicated in Figure 2-8) to find the percent value of the tolerance of this resistor. Insert this value into the Tolerance Percent column of Table 2-1.

3. Minimum and maximum resistance limits:
   a. Use the percent tolerance to find the maximum acceptable resistance limit where:
      Maximum Resistance = Coded Resistance + Percent Tolerance of Coded Resistance
   b. Use the percent tolerance to find the minimum acceptable resistance Limit where:
      Minimum Resistance = Coded Resistance − Percent Tolerance of coded Resistance
   c. Insert the maximum and minimum acceptable resistance limit values into Table 2-1.

4. Repeat this process for all of the remaining resistor values listed in Table 2-1.

NOTE: As stated in the theory and shown in Figure 2-7, the first resistor color code band of a four band general purpose resistor indicates the first digit of the resistance value, the second band indicates the second digit, and the third band is the multiplier (used to determine the number of zeros). The fourth color band indicates tolerance, where red indicates 2%, gold indicates 5%, and silver indicates 10%.

COLOR	DIGIT	MULTIPLIER	TOLERANCE
Black (BK)	0	$10^0 = 1$	
Brown (BR)	1	$10^1 = 10$	$\pm 1\%$
Red (R)	2	$10^2 = 100$	$\pm 2\%$
Orange (O)	3	$10^3 = 1,000$	
Yellow (Y)	4	$10^4 = 10,000$	
Green (G)	5	$10^5 = 100,000$	
Blue (B)	6	$10^6 = 1,000,000$	
Violet (V)	7	$10^7 = 10,000,000$	
Gray (GR)	8	$10^8 = 100,000,000$	
White (W)	9	$10^9 = 1,000,000,000$	
Gold		$10^{-1} = 0.1$	$\pm 5\%$
Silver		$10^{-2} = 0.01$	$\pm 10\%$

TABLE 2-2: Standard resistor color code.

FIGURE 2-7: Standard resistor color code.

PART B: Measuring Resistors

1. With the EWB program running choose open from File menu, select EWB, and (when the dialog box appears) select Chapter 2 and Figure 2-8. Double click on the multimeter, select the ohmmeter by clicking on Ω, set to dc, and measure across each resistor. Because one side of each resistor and the ohmmeter are connected to a common ground point, as shown in Figure 2-8, it is necessary to connect the ohmmeter to only one side of each resistor in making the measurement.

2. Follow these basic steps:
   a. Set the multimeter to dc and ohms.

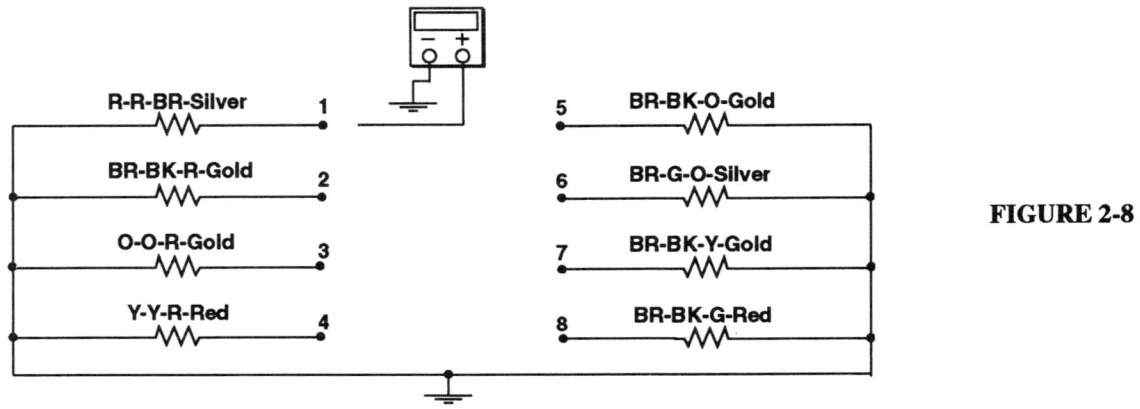

FIGURE 2-8

b. Connect the test lead to terminal 1 to effectively connect the meter across the R-R-BR resistor.

c. Measure the resistance of the resistor under test by clicking on the power switch located in the upper right hand side of the window. A second method is to use the Analysis menu and click on activate.

d. Read the measured value on the meter.

3. Record the measured value of each resistor, as indicated, into Table 2-1.

NOTE: The resistance of resistors are set to slightly different values than the coded values in order to replicate the real world resistance of actual resistors. If the instructions on how to use the program are not clear to you, consult the text material or ask your instructor/professor for assistance.

4. Find the percent tolerance of four (4) of the measured resistors listed in Table 2-1.
   a. Begin with the R-R-BR resistor. If the resistance value is higher than the coded value it does not require a sign, but if the measured resistance value is lower than the coded resistance value indicated, use a minus (−) sign. The percent tolerance, knowing the measured and coded resistance values, is determined from:

   $$\% \text{ Tolerance} = \frac{\text{Measured Resistance} - \text{Coded Resistance}}{\text{Coded Resistance}} \times 100$$

   b. Insert the calculated % tolerance values for each of the four resistors, as indicated, in Table 2-1.

SECTION II: Resistors Connected in Series
PART A: Two Resistors in Series
1. In the two resistor series connection of Figure 2-9(a) and the component diagram of Figure 2-9(b):
   a. Determine the individual coded R1 and R2 resistance values, where R1 is O-O-R and R2 is Y-V-R.
   b. Solve the total resistance of R1 and R2 in series where: $R_T = R1 + R2$.

FIGURE 2-9

2. Connect, or open and select from the File menu, two resistor in series, as shown in Figure 2-10.
   a. Measure the individual R1 and R2 resistor as shown in Figure 2-10(a).
   b. Measure the total series resistance of R1 and R2, as shown in Figure 2-10(b).

3. Insert the calculated and measured values, as indicated, into Table 2-3.

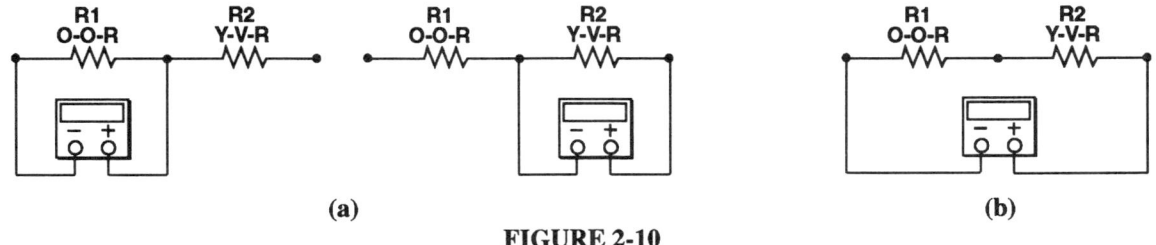

FIGURE 2-10

PART B: Three Resistors in Series
1. With reference to the three resistor connected in series in Figure 2-11:
   a. Determine the R3 resistor value from its BR-BK-O color code. (R1 and R2 were previously decoded.)
   b. Find the total series resistance of the R1, R2, and R3 resistors where: $R_T = R1 + R2 + R3$.
2. Connect, or open and select from the File menu, three resistor in series, as shown in Figure 2-11.
   a. Measure the individual R3 resistance.
   b. Measure the total series resistance of the R1, R2, and R3 circuit.
3. Insert the calculated and measured resistances, as indicated, into Table 2-3.

16 — Basic Circuit Analysis For Electronics Using Electronics Workbench®

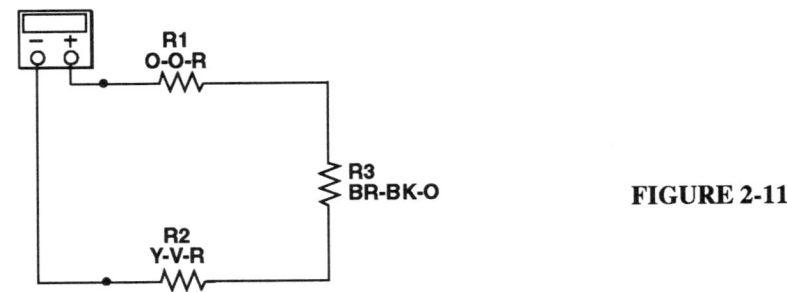

FIGURE 2-11

TABLE 2-3	FIGURES 2-10			FIGURE 2-11	
	R1	R2	$R_T = R1 + R2$	R3	$R_T = R1 + R2 + R3$
CALCULATED					
MEASURED					

SECTION III: Resistors Connected in Parallel

PART A: Two Resistors in Parallel

1. With reference to the two resistor connected in parallel in Figure 2-12:
    a. Determine the individual R1 and R2 resistor values from their color codes, where R1 is O-O-R and R2 is Y-V-R
    b. Solve the resistance of parallel resistors R1 and R2 where: $R_T = R1 \parallel R2 = \dfrac{1}{1/R1 + 1/R2}$.

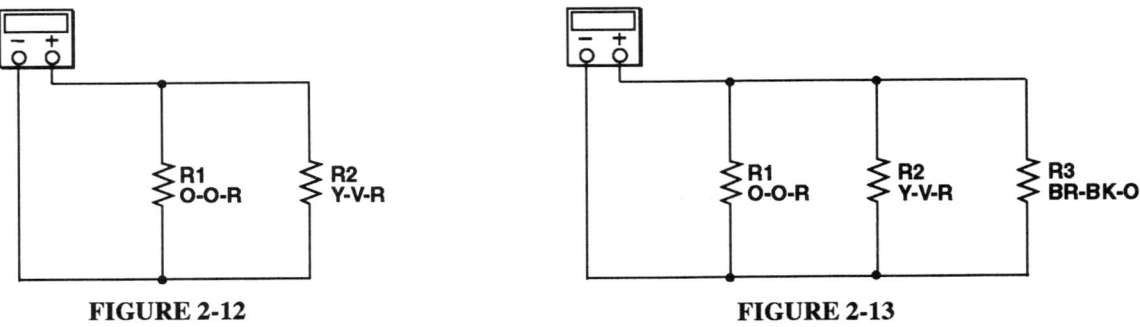

FIGURE 2-12   **FIGURE 2-13**

2. Connect, or open and select from the File menu, two resistors in parallel, as shown in Figure 2-12. Measure the total resistance of the parallel R1 and R2 resistors.

NOTE: Also, two resistors in parallel can be solved from: $R_T = \dfrac{R1 \times R2}{R1 + R2}$.

3. Insert the calculated and measured values, as indicated, into Table 2-4.

PART B: Three Resistors in Parallel

1. With reference to the three resistor parallel connection of Figure 2-13:
    a. Determine the color-coded resistance values of the R1, R2, and R3 resistors where: R1 is O-O-R, R2 is Y-V-R, and R3 is BR-BK-O.
    b. Solve the total resistance of the three resistor parallel connection of Figure 2-13 where:
    $$R_T = R1 \parallel R2 \parallel R3 = \dfrac{1}{1/R1 + 1/R2 + 1/R3}$$

2. Connect, or open and select from the File menu, three resistors in parallel, as shown in Figure 2-13. Measure the total resistance of the parallel R1, R2, and R3 resistors.

3. Insert the calculated and measured value, as indicated, into Table 2-4.

Resistors and Resistance Measurements — 17

TABLE 2-4	FIGURE 2-12			FIGURE 2-13			
	R1	R2	R_T = R1 ∥ R2	R1	R2	R3	R_T
CALCULATED							
MEASURED	///	///		///	///	///	

SECTION IV: Resistors Connected in Series-Parallel

PART A: Three Resistor Series-Parallel Circuit

1. With reference to the series-parallel connection of Figure 2-14:
   a. Determine the color-coded resistance values of each of the three resistors.
   b. Calculate the total resistance of the series-parallel connection where: R_T = R1 + (R2 ∥ R3).

2. Connect, or open and select from the File menu, the series-parallel resistor connection of Figure 2-14. Measure the total resistance of the circuit.

3. Insert the calculated and measured values, as indicated, into Table 2-5.

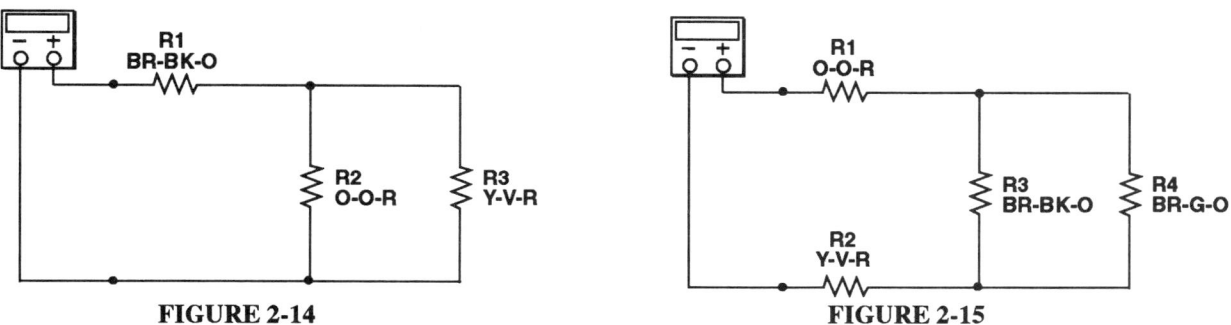

FIGURE 2-14   FIGURE 2-15

PART B: Four Resistor Series-Parallel Connection

1. With reference to the series-parallel connection of Figure 2-15:
   a. Determine the color-coded resistance values of each of the four resistors.
   b. Calculate the total resistance of the series-parallel connection where: R_T = R1 + (R3 ∥ R4) + R2.

2. Connect, or open and select from the File menu, the series-parallel resistor connection of Figure 2-15. Measure the total resistance of the series-parallel resistor connection.

3. Insert the calculated and measured values, as indicated, into Table 2-5.

TABLE 2-5	FIGURE 2-14				FIGURE 2-15				
	R1	R2	R3	R_T	R1	R2	R3	R4	R_T
CALCULATED									
MEASURED	///	///	///		///	///	///	///	

SECTION V (Optional Lab. Exercise): Variable Resistor Measurements

PART A (Guided Lab. Exercise): Determine the resistance across terminals 1 and 3 of the 10 kΩ variable resistor at 10% and 90% rotation as shown in Figure 2-16.

18 — Basic Circuit Analysis For Electronics Using Electronics Workbench®

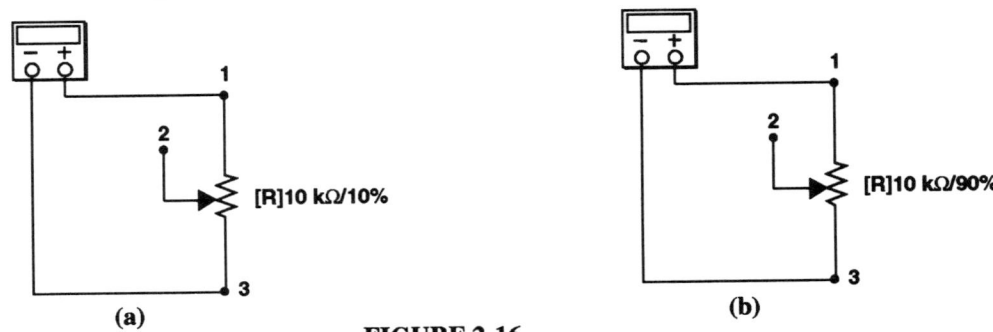

FIGURE 2-16

1. 10% rotation: Connect, or open and select from the File menu, the variable resistor of Figure 2-16(a), where the variable resistance is set at 10% rotation.
   a. Calculate the resistance between terminals 1 and 3 at 10% rotation.
   b. Measure the resistance between terminals 1 and 3 at 10% rotation.

2. 90% rotation: Connect, or open and select from the File menu, the variable resistor of Figure 2-16(b), where the variable resistance is set at 90% rotation.
   a. Calculate the resistance between terminals 1 and 3 at 90% rotation.
   b. Measure the resistance between terminals 1 and 3 at 90% rotation.

3. Insert the calculations and measurements, as indicated, into Table 2-6.

NOTE: Potentiometers are usually constructed so when the shaft is rotated to 90% rotation, the resistance between terminals 2 and 3 is as 90% of maximum. Then, when the shaft is rotated to 10% of full rotation, the resistance between terminals 2 and 3 is at 10% maximum. To change the percent of the variable resistor, double click on the resistor and set the percentage directly. Also, using keys shift and R increases the resistance and using key R alone decreases the resistance.

PART B: (Guided Laboratory): Determine the resistance across terminals 2 and 3 of the 10 kΩ variable resistor at 10% and 90% rotation as shown in Figure 2-17.

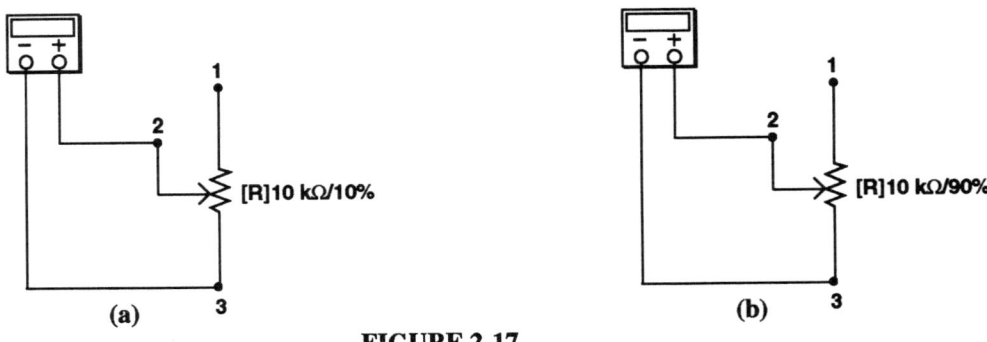

FIGURE 2-17

1. 10% rotation: Connect, or open and select from the File menu, the variable resistor of Figure 2-17(a), where the variable resistance is set at 10% rotation.
   a. Calculate the resistance between terminals 2 and 3 at 10% rotation.
   b. Measure the resistance between terminals 2 and 3 at 10% rotation.

2. 90% rotation: Connect, or open and select from the File menu, the variable resistor of Figure 2-17(b), where the variable resistance is set at 90% rotation.
   a. Calculate the resistance between terminals 2 and 3 at 90% rotation.
   b. Measure the resistance between terminals 2 and 3 at 90% rotation.

3. Insert the calculations and measurements, as indicated, into Table 2-6.

PART C: (Guided Laboratory): Determine the resistance across terminals 1 and 2 of the 10 kΩ variable resistor at 10% and 90% rotation as shown in Figure 2-18.

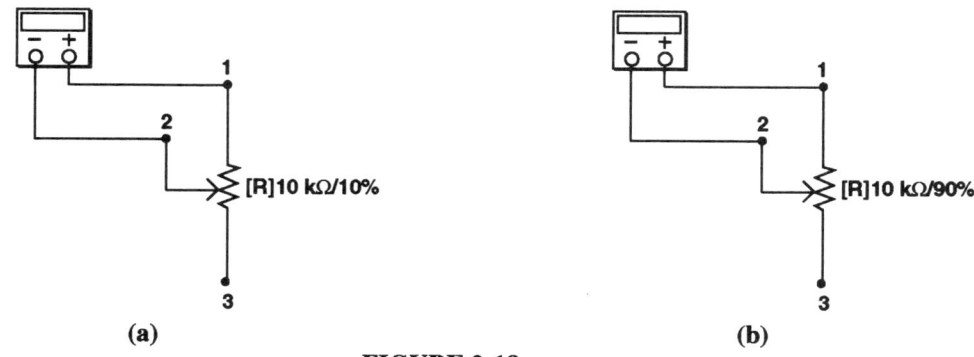

FIGURE 2-18

1. 10% rotation: Connect, or open and select from the File menu, the variable resistor of Figure 2-18(a), where the variable resistance is set at 10% rotation.
   a. Calculate the resistance between terminals 1 and 2 at 10% rotation.
   b. Measure the resistance between terminals 1 and 2 at 10% rotation.

2. 90% rotation: Connect, or open and select from the File menu, the variable resistor of Figure 2-18(b), where the variable resistance is set at 90% rotation.
   a. Calculate the resistance between terminals 1 and 2 at 90% rotation.
   b. Measure the resistance between terminals 1 and 2 at 90% rotation.

3. Insert the calculations and measurements, as indicated, into Table 2-6.

TABLE 2-6	Terminals 1,3		Terminals 2,3		Terminals 1,2	
	10%	90%	10%	90%	10%	90%
VOLTMETER						
MULTIMETER						

Questions and Problems: Basic Circuit Analysis for Electronics: 38-40.

CHAPTER 3
OHM'S LAW

INTRODUCTION

The discovery of the mathematical relationship between current, voltage, and resistance in an electric circuit by Georg Ohm, in the early 1800's, marks the beginning of the practical use of electricity. What Ohm discovered was that current in an electric circuit is directly proportional to the applied voltage and inversely proportional to the resistance. In other words, Ohm's law states that current in amperes in an electric circuit is equal to the voltage in volts divided by the resistance in ohms. Using algebraic manipulations, the other forms of Ohm's law are $V = IR$ and $R = V/I$. With Ohm's law it became possible to design circuits to transmit electrical energy to devices and machines that convert electrical energy to other types of energy (heat, magnetic, chemical, etc.) to do functional work.

DIRECT CURRENT VOLTAGE

The dc voltage source used in Electronics Workbench® is the battery and its symbol is shown in Figure 3-1(a). To obtain and place the dc battery on the workspace, click on the Sources bin and move the arrow to the battery. Click on the mouse button, continue to hold the button down, and drag the battery into the workspace. Set the battery in the desired position and release the button.

The battery has a default voltage of 12 V. To change the voltage, double click on the battery symbol. When the dialog box appears select Value, type in the desired value, say 18 V, and click on OK. To label the power supply, select label, type in V_{PS}, and click on OK. The 18 V and V_{PS} will appear, as shown in Figure 3-1(b). To move additional test equipment or components to or around the workspace, click on the item and once it is highlighted use the mouse to drag and place it where wanted.

FIGURE 3-1

VOLTMETERS

Voltage is typically measured on the multimeter obtained by dragging it from the Instruments bin to the workspace. Multimeters have the capability of measuring volts, amperes, and ohms. So, in order to select dc volts, double click on the multimeter icon to make it larger and set to volts (V) and dc (—), as shown in Figure 3-2(a). To change or check the settings, such as the internal resistance of the voltmeter, click on the settings. When the dialog box appears, change the resistance setting by typing in the required value. Then, to measure the voltage, connect the voltmeter across the power supply or circuit component and click on the power switch to run the simulation. The measured voltage is displayed on the meter.

Another voltmeter used in EWB, found in the Indicators bin on the Part Bin toolbar, is shown in Figure 3-2(b). It has a default voltage setting, but to check or change the mode click on the meter, click on the mode, select dc (or ac), and click on OK. Since the dc voltage measurement requires meter polarity to be observed correctly, one side of the meter has a darker line to indicate the negative connection. An incorrect polarity connection (positive and negative leads reversed) will produce a negative (–) sign preceding the numerical readout on the meter.

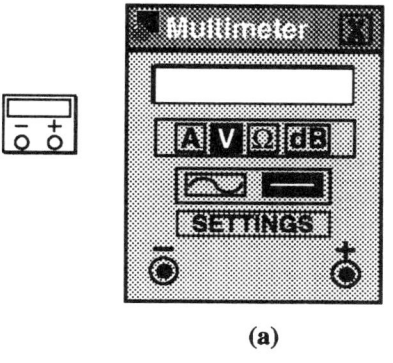

(a)

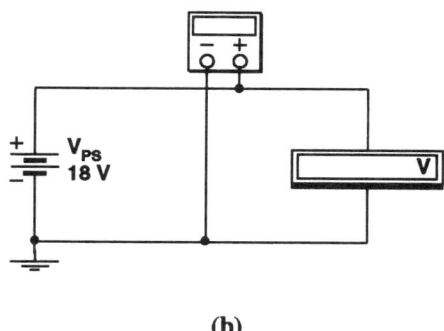

(b)

FIGURE 3-2

AMMETERS

Ammeters have very low internal resistance and are always connected in series with the circuit resistors. So, if an ammeter is connected directly across a voltage source, it has the same effect as placing a short across the voltage source. In hardware laboratories it would produce a high current that would blow the protective fuses and, if the fuses were not fast enough, it could damage both the meter and the circuit under test.

On the Electronics Workbench® workspace, the ammeter is obtained by using the multimeter set to (A) and dc (—), as shown in Figure 3-3(a). To change or check the setting, such as the internal resistance of the ammeter, double click on the settings. When the dialog box appears, change the resistance setting by typing in the required values.

Another ammeter used in EWB, found in the Indicators bin, is shown in Figure 3-3(b). This ammeter has a default dc setting. To change or check the mode, double click on meter, click on mode, select dc (or ac), and click OK. For dc circuit measurements, the polarities of the ammeter, like those of the voltmeter, must be observed. The darker line on the meter indicates the negative side. An incorrect connection (where the positive and negative leads are mistakenly reversed) will produce a negative (–) sign displayed on the meter readout.

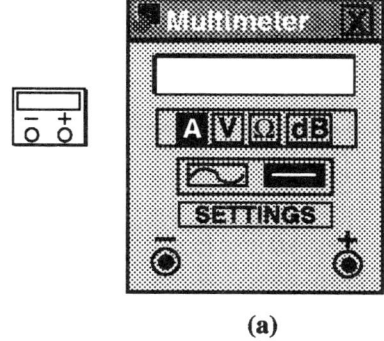

(a)

(b)

FIGURE 3-3

THE LABORATORY EXERCISE

The laboratory exercise is divided into three sections. Each section has both pre-laboratory calculations and measurements.

In Section I of the laboratory exercise, both the multimeter, set to dc volts, and the voltmeter are used to measure positive and negative voltages. Then, the multimeter and the voltmeter are compared for accuracy.

In Section II, an ammeter is connected and used to measure current through a fixed value resistor while varying the degrees of applied voltage.

In Section III, unknown resistor values are obtained by using a fixed voltage and an ammeter to solve resistance using the formula $R = V/I$. The resistance value is then verified by disconnecting the resistor from the circuit and using the ohmmeter to directly measure the resistance.

LABORATORY EXERCISE

READING ASSIGNMENT: Basic Circuit Analysis for Electronics: 41-47.

EXERCISE OBJECTIVES
To become familiar with:

- Basic voltage and current measurements using voltmeters and ammeters.
- Verifying Ohm's law using voltage and current measurements.

PROCEDURE

SECTION I (Guided Laboratory Exercise): Using Voltmeters
PART A: Measuring the Power Supply Voltage
In the connection of Figure 3-4(a), the power supply (battery) is set to 12 V and measured with both the voltmeter and the multimeter set on volts and dc. Then the battery is reversed, as shown in the connection of Figure 3-4(b), so that the voltage measured is a negative voltage.

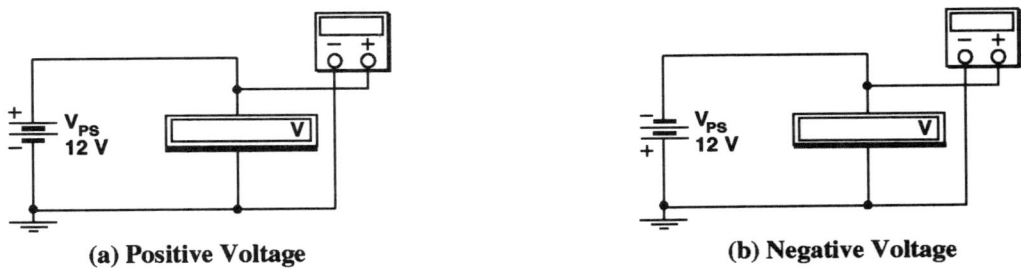

(a) Positive Voltage   (b) Negative Voltage

FIGURE 3-4

NOTE: The wire connected from the positive terminal of the power supply to the positive terminal of the voltmeter is typically red, and the wire connected to the negative terminal of the voltmeter is typically black. To change the voltage value, double click on the voltage (battery icon), change the value and click accept. Try 18 V for example. Then, switch back to 12 V using the same procedure. Also, to change the color of a connecting wire, double click on the wire and, when the color choices appears, select a color.

1. Voltmeter Measurements:
   a. Connect, or open and select from the File menu, the circuit of Figure 3-4(a), where the power supply (battery) is set to 12 V. Measure V_{PS} with the voltmeter and the multimeter (set on volts and dc).
   b. Connect, or open and select from the File menu, the circuit of Figure 3-4(b), where the power supply V_{PS} is reversed. Measure the negative voltage using both the voltmeter and multimeter.

2. Insert both the 12 V and the –12 V measured dc voltage values, as indicated, into Table 3-1.

PART B: Achieving the Best Accuracy.
Voltage can be measured on more than one meter, with varying degrees of accuracy, depending on whether the voltage to be measured is below or above 10 V and on what meter is used. Therefore, for the purpose of voltmeter familiarity, measure the two voltages below 10 V (9.1416 V and 9.1515 V) as shown in Figure

3-5(a), with both the voltmeter and the multimeter set on volts and dc. Then measure the two voltages above 10 V (12.145 V and 12.156 V), as shown in Figure 3-5(b), with both the voltmeter and the multimeter set on volts and dc.

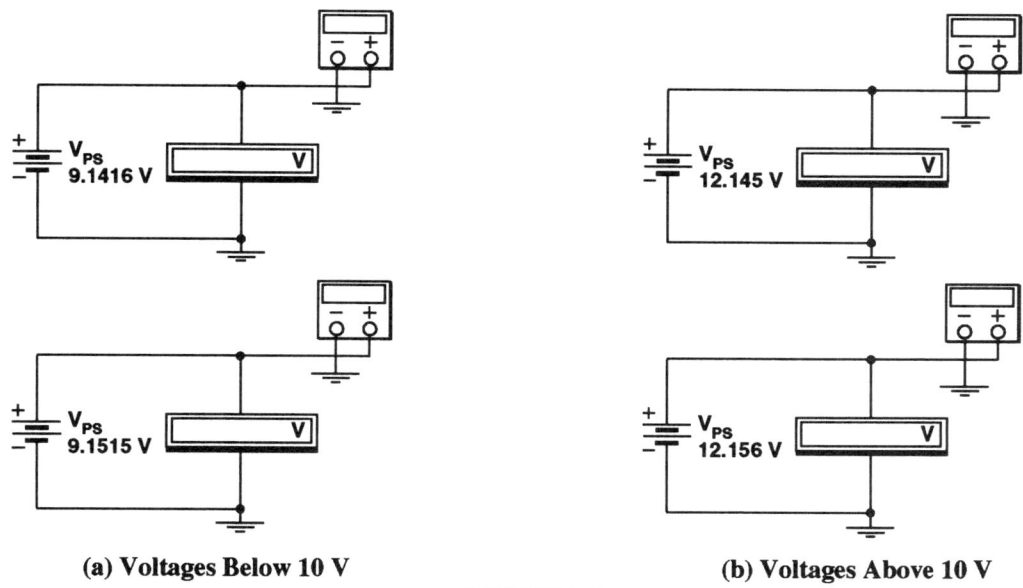

(a) Voltages Below 10 V (b) Voltages Above 10 V

FIGURE 3-5

1. Power supply voltage measurements (less than 10 V):
   a. Connect, or open and select from the File menu, the circuits shown in Figure 3-5.
   b. If the circuit is being constructed and not selected, double click on the power supply (battery) and set the power supply voltage initially to 9.1416 V, as shown in Figure 3-5(a).
   c. Measure the voltage on the voltmeter and on the multimeter with both set on volts and dc.
   d. Repeat the voltage measurements with the power supply voltage set to 9.1515 V.
   e. Insert the measured voltmeter and multimeter voltages (< 10 V), as indicated, into Table 3-1.

2. Power supply voltage measurements (greater than 10 V):
   a. Initially the power supply voltage is set to 12.145 V, as shown in Figure 3-5(b).
   b. If the circuit is being constructed and not selected, double click on the power supply (battery) and set the power supply voltage to 12.145 V.
   c. Measure the voltage on the voltmeter and on the multimeter set on volts and dc.
   d. Repeat the voltage measurements with the power supply voltage set to 12.156 V.
   e. Insert the measured voltmeter and multimeter voltages (> 10 V), as indicated, into Table 3-1.

TABLE 3-1	FIGURE 3-4		FIGURE 3-5(a)		FIGURE 3-5(b)	
	12 V	−12 V	9.1416 V	9.1515 V	12.145 V	12.156 V
VOLTMETER						
MULTIMETER						

SECTION II (Guided Laboratory Exercise): Measuring Current
PART A: Pre-laboratory Calculations

In the circuits of Figure 3-6, a 1.5 kΩ resistor is added to the circuit and the power supply is varied from (initially) 6 V, then to 12 V, and lastly to 18 V. At each voltage of the voltage conditions, solve the current through the 1.5 kΩ resistor and the power developed by the resistor.

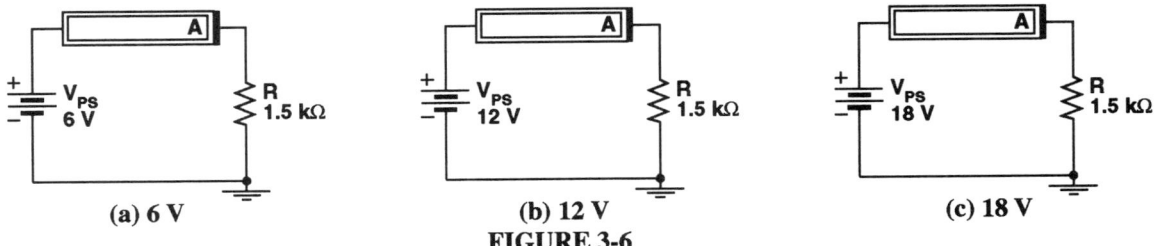

FIGURE 3-6

1. 6 V power supply calculations:
   a. Calculate the current through the 1.5 kΩ resistor of Figure 3-6(a) where: $I = V/R$.
   b. Calculate the power dissipated by the resistor where: $P = VI = I^2R = V^2/R$.

2. 12 V power supply calculations:
   a. Calculate the current through the 1.5 kΩ resistor of Figure 3-6(b) where: $I = V/R$.
   b. Calculate the power dissipated by the resistor where: $P = VI = I^2R = V^2/R$.

3. 18 V power supply calculations:
   a. Calculate the current through the 1.5 kΩ resistor of Figure 3-6(c) where: $I = V/R$.
   b. Calculate the power dissipated by the resistor, where $P = VI = I^2R = V^2/R$.

4. Insert the calculated current and power for each of the voltage conditions, as indicated, into Table 3-2.

PART B: Measuring Current

Connect, or open and select from the File menu, the circuits shown in Figure 3-6. In the connection of Figure 3-6(a), the power supply (battery) is set to 6 V and the current is measured with the ammeter. Then this procedure is repeated for 12 V, as shown in Figure 3-6(b), and 18 V, as shown in Figure 3-6(c).

1. Activate the circuit and measure the respective voltages and currents for each of the Figure 3-6 circuits.

2. Insert the calculated and measured values, as indicated, into Table 3-2.

TABLE 3-2	V	I	P	V	I	P	V	I	P
CALC.	6 V			12 V			18 V		
MEAS.			/////			/////			/////

SECTION III: Solving for an Unknown Resistance when the Voltage and Current are Known

Connect, or open and select from the File menu, the circuit shown in Figure 3-7, where three (3) resistors with unknown values are substituted in the circuit for R. Each of the three resistors will have different multiplier values of red (10^2), orange (10^3), and yellow (10^4). The resistor values will be solved (calculated) using a fixed power supply voltage of 18 volts and the measured current through each resistor under test.

1. Find the unknown resistance value of a resistor with a red colored multiplier band.
   a. The first unknown resistor has a red (2) multiplier band. Since the multiplier band indicates a resistance value that should be in the 1 kΩ to 10 kΩ region, the current flow through the resistor should be in the 1.2 mA to 12 mA range.
   b. Measure the current for the unknown red multiplier banded resistor, as shown in Figure 3-7(a).

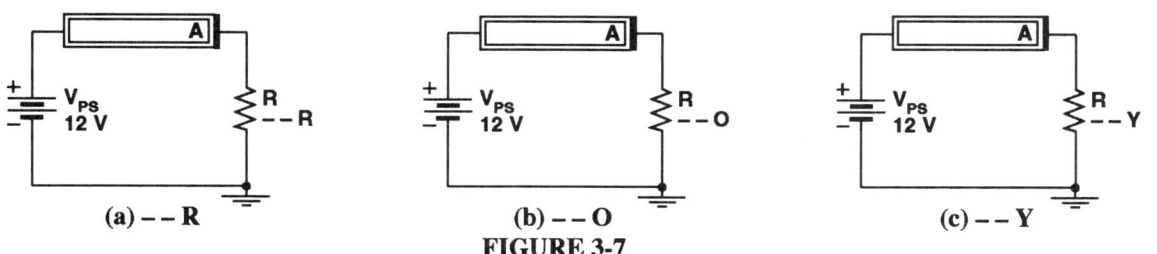

FIGURE 3-7

c. Calculate the unknown resistance value from $R = V/I$ where: $V = 12$ V and I is measured.
d. Determine the expected nominal resistance.
e. Identify the expected color code of the resistor.
f. Verify the resistance value by measuring the resistor with an ohmmeter.

NOTE: Resistor measurements with an ohmmeter should be made with the resistor disconnected from the circuit so the resistor being measured is free from any circuit voltages.

2. Repeat the Step 2 procedure for a second resistor with an orange (3) multiplier band, as shown in Figure 3-7(b). Since the multiplier band indicates a resistance value that should be in the 10 kΩ to 100 kΩ region, the current flow through the resistor should be in the 120 μA to 1.2 mA range. Include the measured ohmmeter resistance, the expected nominal resistance, and the color code of the resistor under test.

3. Repeat the Step 2 procedure for a third resistor with a yellow (4) multiplier band, as shown in Figure 3-7(c). Since the multiplier band indicates a resistance value that should be in the 100 kΩ to 1 MΩ region, the current flow through the resistor should be the 12 μA to 120 μA range. Include the measured ohmmeter resistance, the expected nominal resistance, and the color code of the resistor under test.

4. Insert the calculated and measured values, as indicated, into Table 3-3.

TABLE 3-3	R		O		Y	
	I_{R1}	R1	I_{R2}	R2	I_{R3}	R3
CALCULATED						
MEASURED						
EXPECTED NOMINAL RESISTANCE						
VERIFICATION USING THE OHMMETER						

Questions and Problems: Basic Circuit Analysis for Electronics: 52-54.

CHAPTER 4
SERIES CIRCUITS

INTRODUCTION

The analysis of two resistors in series connected to a power supply can be solved either by Ohm's law or voltage divider equations. In using Ohm's law to analyze the two resistor circuit of Figure 4-1(a), the total resistance R_T is found where: $R_T = R1 + R2$. Then, the current through R1 and R2 is found from: $I_T = V_{PS}/R_T$. Once the current is known, V_{R1} is found by multiplying the current by the R1 resistance and V_{R2} is found by multiplying the currently by the R2 resistance.

When voltage divider equations are used to analyze the circuit of Figure 4-1(a), the voltage drops can be solved without having to solve R_T and the circuit current I_T first. The equations are shown in conjunction with Figure 4-1(b) where $V_{R1} = V_{PS} \times R1/(R1 + R2)$ and $V_{R2} = V_{PS} \times R2/(R1 + R2)$.

Three resistors in series in a circuit are solved much like the two resistor by expanding the equations to cover the additional resistor. The Ohm's law method is shown in conjunction with the circuit of Figure 4-1(c) and the voltage divider method is shown in conjunction with Figure 4-1(d).

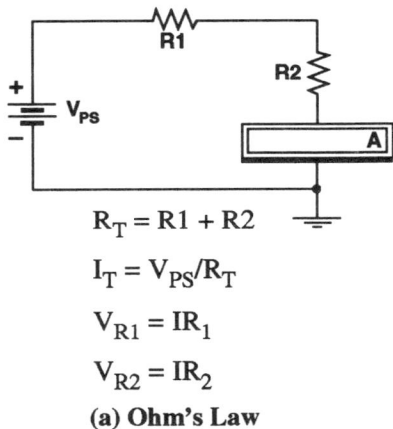

$R_T = R1 + R2$

$I_T = V_{PS}/R_T$

$V_{R1} = IR_1$

$V_{R2} = IR_2$

(a) Ohm's Law

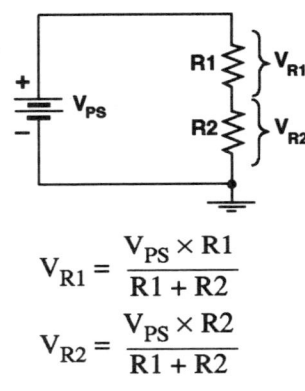

$V_{R1} = \dfrac{V_{PS} \times R1}{R1 + R2}$

$V_{R2} = \dfrac{V_{PS} \times R2}{R1 + R2}$

(b) Voltage Divider Equations

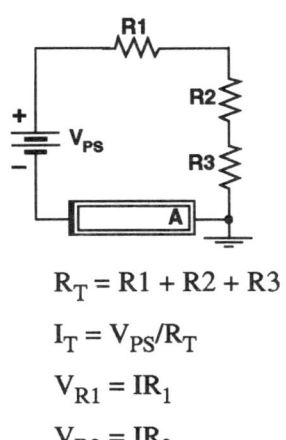

$R_T = R1 + R2 + R3$

$I_T = V_{PS}/R_T$

$V_{R1} = IR_1$

$V_{R2} = IR_2$

$V_{R3} = IR_3$

(c) Ohm's Law

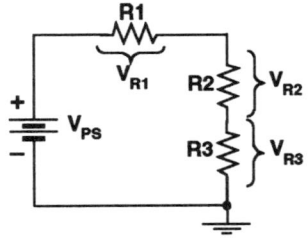

$V_{R1} = \dfrac{V_{PS} \times R1}{R1 + R2 + R3}$

$V_{R2} = \dfrac{V_{PS} \times R2}{R1 + R2 + R3}$

$V_{R3} = \dfrac{V_{PS} \times R3}{R1 + R2 + R3}$

(d) Voltage Divider Equations

FIGURE 4-1

LABORATORY EXERCISE

READING ASSIGNMENT: Basic Circuit Analysis for Electronics: 55-60

EXERCISE OBJECTIVES
To become familiar with:
- Using Ohm's law to determine the current and voltages in a series circuit.
- Using voltage divider equations to determine the voltage drops in series circuits.
- Troubleshooting opens and shorts in series circuits.

PROCEDURE

SECTION I (Guided Lab. Exercise): Two Resistor Series Circuit

PART A: Pre-laboratory Calculations
Calculate voltage drops and circuit current of the two resistor circuit of Figure 4-2. In the calculations, the V_{PS} is 18 V and the series resistances are R1 at 10 kΩ and R2 at 15 kΩ.

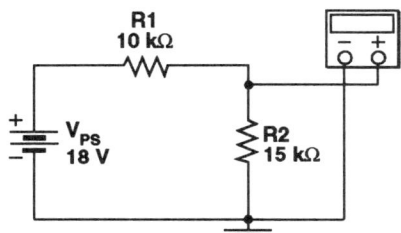

FIGURE 4-2

1. METHOD 1: Solve V_{R1} and V_{R2}. Use Ohm's law to solve R_T, I, and then V_{R1} and V_{R2}.
   a. Calculate R_T, the total series resistance, where: $R_T = R1 + R2$
   b. Calculate I, the series circuit current, where: $I = V_{PS}/R_T$.
   c. Calculate the voltage drops across the series resistors R1 and R2, where: $V_{R1} = IR_1$ and $V_{R2} = IR_2$.

2. METHOD 2: Solving V_{R1} and V_{R2} using the voltage divider equations.

   a. $V_{R1} = \dfrac{V_{PS} \times R1}{R1 + R2}$ b. $V_{R2} = \dfrac{V_{PS} \times R2}{R1 + R2}$

3. Insert the calculated values, as indicated, into Table 4-1.

PART B: Voltage Measurements
Construct, or select from the File menu, the circuit of Figure 4-2. In the connection, the power supply is set to 18 V and measured with the multimeter set on volts (V) and dc (—).

1. Measure V_{R2}. Measure directly across R2 with respect to reference ground.

2. Find V_{R1} indirectly. Use the measured V_{PS} and V_{R2} voltages and determine V_{R1} where: $V_{R1} = V_{PS} - V_{R2}$.

28 — Basic Circuit Analysis For Electronics Using Electronics Workbench®

NOTE: In order to promote good measuring techniques, all voltage measurements are typically made with reference to a common point in the circuit. For this circuit, the common or reference ground is the negative terminal of the power supply. In practice, the possibility of more than one ground, such as when both the circuit and test equipment are separately grounded can produce a ground loop. Measuring with respect to a common point in the circuit helps avoid that possibility.

3. Insert the measured values into Table 4-1.

PART C: Current Measurements

1. Connect or select from the File menu the circuit of Figure 4-3. Measure the current using either the direct or indirect methods.
    a. In the direct method, shown in Figure 4-3, current is measured by inserting the ammeter in series with the circuit resistors.
    b. In the indirect method, use the measured V_{R2} voltage divided by the R2 resistance value.

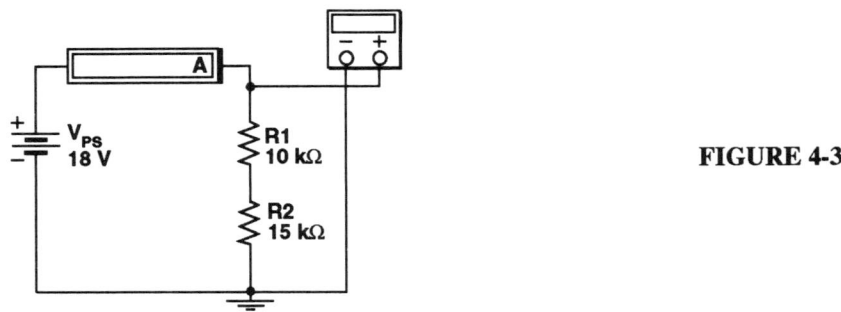

FIGURE 4-3

NOTE: The internal resistance of an ammeter is ideally zero ohms. So make sure the ammeter is connected in series with the circuit resistors and that the polarity is observed.

2. Insert the measured values, as indicated, into Table 4-1.

TABLE 4-1	R_T	I	V_{R1}	V_{R2}	I_{R1}	I_{R2}
CALCULATED						
MEASURED						

SECTION II: Three Resistors in Series

PART A: Pre-laboratory Calculations

Calculate voltage drops and circuit current of the three resistor circuit of Figure 4-4. In the calculations, use the V_{PS} at 18 V and the series resistance values of R1 = 12 kΩ, R2 = 4.7 kΩ, and R3 = 3.3 kΩ.

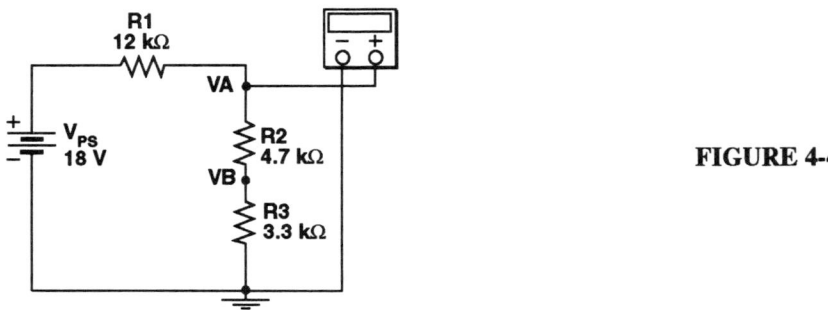

FIGURE 4-4

Series Circuits — 29

1. METHOD 1: Solve V_{R1}, V_{R2} and V_{R3} using Ohm's law. Solve R_T, I, and then V_{R1}, V_{R2}, and V_{R3}.
   a. Calculate R_T, the total series resistance, where $R_T = R1 + R2 + R3$
   b. Calculate I, the series circuit current, where $I = V_{PS}/R_T$.
   c. Calculate the voltage drops across the series resistors where: $V_{R1} = IR_1$, $V_{R2} = IR_2$, and $V_{R3} = IR_3$.

2. METHOD 2: Solve V_{R1}, V_{R2} and V_{R3} using voltage divider equations.

   a. $V_{R1} = \dfrac{V_{PS} \times R1}{R1 + R2 + R3}$ b. $V_{R2} = \dfrac{V_{PS} \times R2}{R1 + R2 + R3}$ c. $V_{R3} = \dfrac{V_{PS} \times R3}{R1 + R2 + R3}$

3. Insert the calculated values, as indicated, into Table 4-2.

PART B: Voltage Measurements

Connect, or open and select from the File menu, the circuit of Figure 4-4. In the connection, the power supply is set to 18 V and the measurements are made with the multimeter set on volts (V) and dc (—).

1. Measure V_{PS} and VA, both measured with respect to reference ground. Then use the measured V_{PS} and VA to determine V_{R1} where: $V_{R1} = V_{PS} - VA$ and $VA = V_{R2} + V_{R3}$.

2. Find V_{R2} indirectly. Measure $VB = V_{R3}$ directly, measured with respect to ground. Then use the measured VB voltages to determine V_{R2} where: $V_{R2} = VA - VB$.

NOTE: Again in order to promote good measuring techniques, all voltage measurements are made with reference to a common point in the circuit to help avoid ground loops.

3. Insert the measured values, as indicated, into Table 4-2.

PART C: Current Measurements

1. Connect, or open and select from the File menu, the circuit of Figure 4-5. Measure the current using either the direct or indirect methods.
   a. In the direct method the current is found by inserting the ammeter in series with the circuit resistors, as shown in Figure 4-5.
   b. In the indirect method the current is found by using the measured V_{R3} voltage divided by the R3 resistance value.

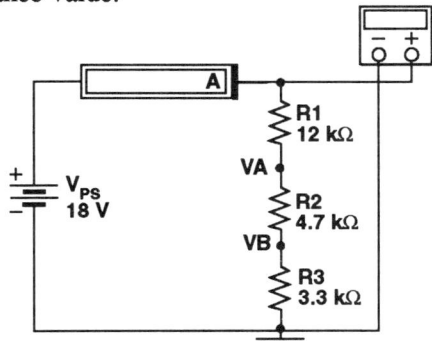

FIGURE 4-5

NOTE: The internal resistance of an ammeter is ideally zero ohms. So the ammeter must be connected in series with the circuit resistance and the polarity of the meter must be observed.

2. Insert the measured values, as indicated, into Table 4-2.

TABLE 4-2	R_T	I	V_{R1}	V_{R2}	$V_B = V_{R3}$	$V_A = V_{(R2 + R3)}$	I_{R3}
CALCULATED						////	////
MEASURED	////						

30 — Basic Circuit Analysis For Electronics Using Electronics Workbench®

SECTION III: Troubleshooting Opens and Shorts

Opens and shorts are common electric circuit problems. An open circuit occurs when parts of the circuit do not make contact. A short occurs when a very low resistance path, such as two bare wires touching, causes a part or parts of the circuit to be bypassed. Both conditions cause the circuit to fail in its proper operation.

In beginning laboratory classes, problems with opens and shorts can occur because connections are not made or resistors are connected across common connection points. This happens because of a lack of familiarity with the circuit layout. Once the equipment is understood and circuit layout becomes familiar the problems disappear and rarely occur again.

PART A: Open Circuit Conditions

1. An open exists between resistors R1 and R2 in the circuit of Figure 4-6.

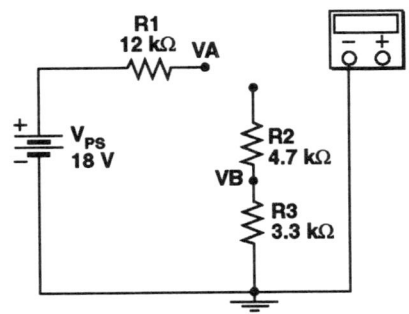

FIGURE 4-6

2. Based on the open condition of Figure 4-6:
   a. Calculate the voltage drop across resistors R1, R2, and R3.
   b. Calculate the current through circuit resistors.
   c. Calculate the expected VA and VB voltages, both with respect to ground.

3. Construct, or open and select from the File menu, the circuit shown in Figure 4-6.
   a. Measure V_{PS}, VA, and VB, measured with respect to ground
   b. Find the measured voltage drops of V_{R1}, V_{R2}, and V_{R3}.
   c. Measure the circuit current, directly or indirectly.

4. Insert the calculated and measured values, as indicated, into Table 4-3.

TABLE 4-3	V_{PS}	V_A	V_B	V_{R1}	V_{R2}	V_{R3}	I
CALCULATED							
MEASURED							

PART B: Short Circuit Conditions

1. A short exists across resistor R2, shorted by a length of wire, as shown in the circuit of Figure 4-7.

2. Based on the shorted condition of the circuit of Figure 4-7:
   a. Calculate the voltage drop across resistors R1, R2, and R3
   b. Calculate the current through the circuit resistors
   c. Calculate the expected VA and VB voltages, both with respect to ground.

3. Construct, or open and select from the File menu, the circuit shown in Figure 4-7.

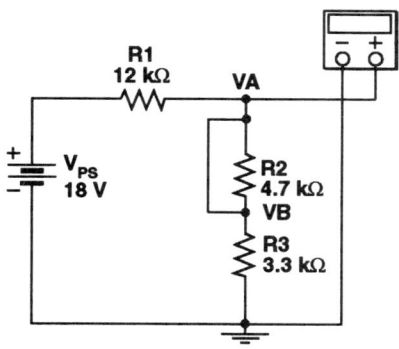

FIGURE 4-7

a. Measure the V_{PS}, VA and VB, measured with respect to ground
b. Find the measured voltage drops of V_{R1}, V_{R2}, and V_{R3}.
c. Measure the circuit current through resistsors R1 and R3, directly or indirectly.

4. Insert the calculated and measured values, as indicated, into Table 4-4

TABLE 4-4	V_{PS}	V_A	V_B	V_{R1}	V_{R2}	V_{R3}	$I_{R1} = I_{R3}$	I_{R2}
CALCULATED	/////							
MEASURED								

Questions and Problems: Basic Circuit Analysis for Electronics: 66-67

CHAPTER 5
PARALLEL CIRCUITS

INTRODUCTION

The analysis of two resistors in parallel connected across a power supply can be solved either by Ohm's law equations or by current divider equations. In the Ohm's law method of solution, shown in conjunction with the circuit of Figure 5-1(a), the currents through the R1 and the R2 resistors are solved individually. Then, the total current is solved by adding the I_{R1} and I_{R2} currents. With the total current known, the total resistance of the circuit is solved by dividing the power supply voltage by the total current where: $R_T = V_{PS}/I_T$. The parallel resistance can also be solved from: $R_T = R1 \| R2$.

A second method for solving the current through each resistor is by use of current divider equations, as shown in conjunction with the circuit of Figure 5-1(a). Thus, when I_T is known, I_{R1} is solved from: $I_{R1} = I_T \times R2/(R1 + R2)$ and I_{R2} is solved from: $I_{R2} = I_T \times R1/(R1 + R2)$.

Three resistors in parallel across a power supply are solved in a similar manner by expanding the equations to include the additional resistor. The Ohm's law method and the current divider method are shown in conjunction with the circuit of Figure 5-1(b).

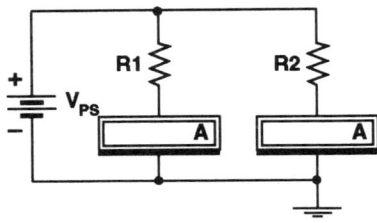
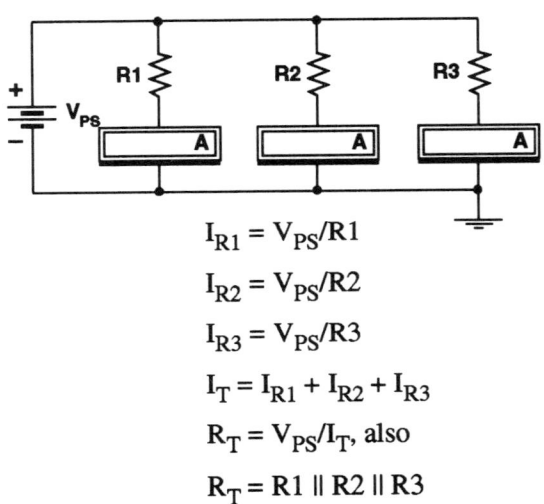

$I_{R1} = V_{PS}/R1$

$I_{R2} = V_{PS}/R2$

$I_T = I_{R1} + I_{R2}$

$R_T = V_{PS}/I_T$, also

$R_T = R1 \| R2$

Ohm's Law

$I_{R1} = V_{PS}/R1$

$I_{R2} = V_{PS}/R2$

$I_{R3} = V_{PS}/R3$

$I_T = I_{R1} + I_{R2} + I_{R3}$

$R_T = V_{PS}/I_T$, also

$R_T = R1 \| R2 \| R3$

Ohm's Law

$$I_{R1} = \frac{I_T \times 1/R1}{1/R1 + 1/R2} = \frac{I_T \times R2}{R1 + R2}$$

$$I_{R2} = \frac{I_T \times 1/R2}{1/R1 + 1/R2} = \frac{I_T \times R1}{R1 + R2}$$

Current Dividers

$$I_{R1} = \frac{I_T \times 1/R1}{1/R1 + 1/R2 + 1/R3}$$

$$I_{R2} = \frac{I_T \times 1/R2}{1/R1 + 1/R2 + 1/R3}$$

$$I_{R3} = \frac{I_T \times 1/R3}{1/R1 + 1/R2 + 1/R3}$$

Current Dividers

(a) Two Resistors in Parallel **(b) Three Resistors in Parallel**

FIGURE 5-1

Parallel Circuits — 33

LABORATORY EXERCISE

READING ASSIGNMENT: Basic Circuit Analysis for Electronics: 68-72

EXERCISE OBJECTIVES
To become familiar with:

- Using Ohm's law to determine the currents in parallel circuits.

- Using Kirchhoff's current law to determine the total current delivered to parallel circuits.

- Using current divider equations to determine the currents in parallel circuits.

PROCEDURE

SECTION 1 (Guided Lab. Exercise): Two Resistors in Parallel
PART A: Pre-laboratory Calculations
Analyze the two resistor parallel circuit of Figure 5-2. In the calculations, the power supply voltage is 18 V and the parallel resistances are R1 at 10 kΩ and R2 at 15 kΩ.

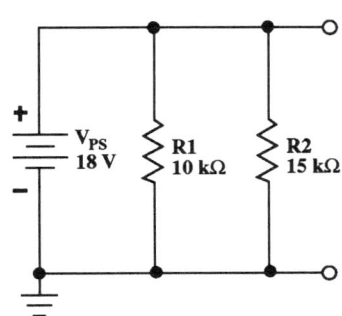 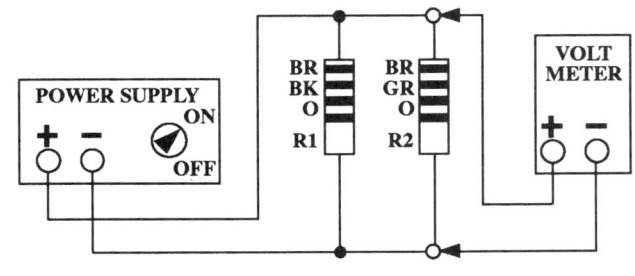

FIGURE 5-2

Two methods can be used to find I_{R1}, I_{R2}, and I_T. In the first method Ohm's law and Kirchhoff's current laws are used, and in the second method Ohm's law and the current divider equations are used. Use either method in the analysis.

1. Method 1: Use Ohm's law to determine I_{R1} and I_{R2}, and Kirchhoff's law to find I_T.
   a. Calculate I_{R1}, the current flow through resistor R1, where: $I_{R1} = V_{PS}/R1$.
   b. Calculate I_{R2}, the current flow through resistor R2, where: $I_{R2} = V_{PS}/R2$.
   c. Calculate I_T, the total output current delivered by the power supply to parallel resistors R1 and R2, where: $I_T = I_{R1} + I_{R2}$.
   d. Calculate R_T, the resistance of R1 and R2 in parallel, where: $R_T = V_{PS}/I_T$.

2. Method 2: Use Ohm's law and the current divider equations to find I_{R1} and I_{R2}.
   a. Calculate R_T, the total parallel resistance, where: $R_T = R1 \parallel R2$.
   b. Calculate I_T, the total current delivered to parallel R1 and R2 resistors, where: $I_T = V_{PS}/R_T$.
   c. Calculate I_{R1}, the current flow through resistor R1, where: $I_{R1} = I_T \times R2/(R1 + R2)$.

d. Calculate I_{R2}, the current flow through resistor R2 where: $I_{R2} = I_T \times R1/ (R1 + R2)$.

3. Insert the calculated values, as indicated, into Table 5-1.

PART B: Current Measurements

1. Voltage Source:
    a. Construct, or select from the File menu, the circuit of Figure 5-3, where the V_{PS} is set to 18 V and the R1 = 10 kΩ and R2 = 15 kΩ resistors are connected in parallel.
    b. Once the circuit is activated, measure the I_{R1} and I_{R2} currents directly, as shown in Figure 5-3(a).
    c. Measure total output current from the power supply where $I_T = I_{R1} + I_{R2}$, as shown in the circuit of Figure 5-3(b).

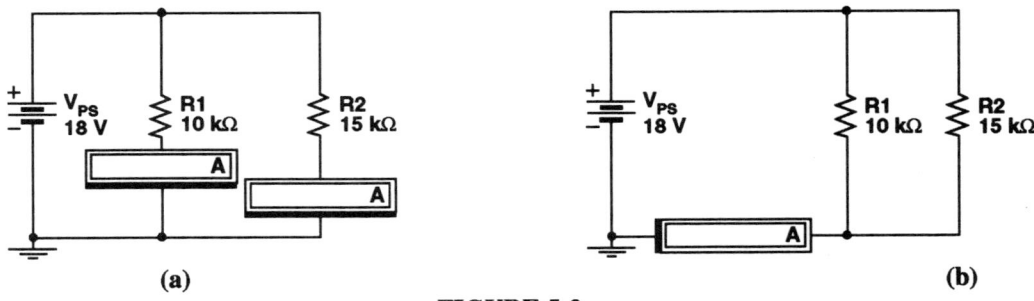

FIGURE 5-3

2. Constant Current Source:
    a. Construct, or select from the File menu, the circuit of Figure 5-4, where the constant current source is set to 3 mA.
    b. Once the circuit is activated, measure the I_{R1} and I_{R2} currents directly, as shown in Figure 5-4(a).
    c. Measure the total output current where $I_T = I_{R1} + I_{R2}$, as shown in the circuit of Figure 5-4(b).

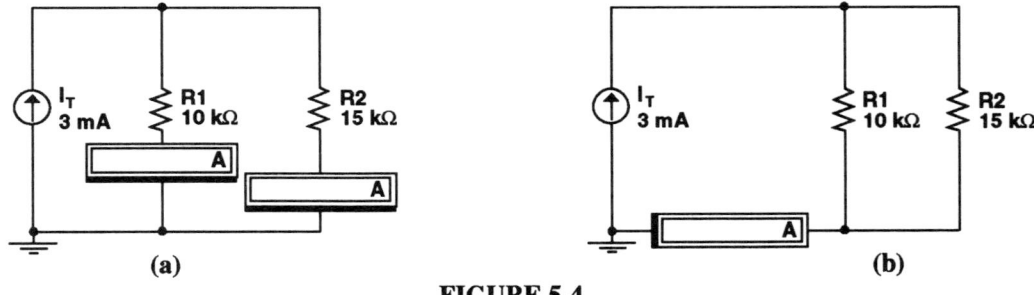

FIGURE 5-4

3. Insert the measured values, as indicated, into Table 5-1.

TABLE 5-1	Voltage Source			Current Source		
	I_{R1}	I_{R2}	I_T	I_{R1}	I_{R2}	I_T
CALCULATED						
MEASURED						

SECTION II: Three Resistors in Parallel

PART A: Pre-laboratory Calculations

Analyze the three resistor parallel circuit of Figure 5-5. In the calculations, the power supply voltage is set to 18 V and the parallel resistances are R1 at 10 kΩ, R2 at 12 kΩ, and R3 at 15 kΩ.

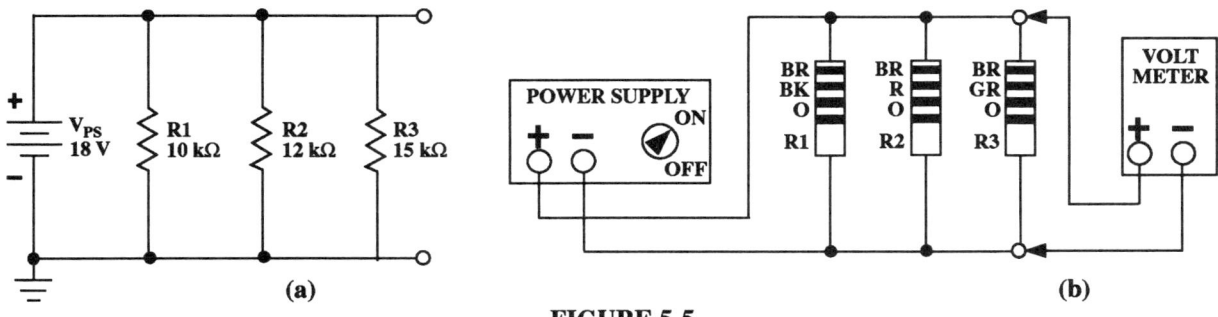

FIGURE 5-5

1. Two methods can be used to find I_{R1}, I_{R2}, and I_{R3}. In the first method Ohm's law and Kirchhoff's current law are used, and in the second method Ohm's law and the current divider equations are used. Use either method in the analysis.

2. Insert the calculated values into Table 5-2.

PART B: Current Measurements

1. Voltage Source:
   a. Construct, or select from the File menu, the circuit of Figure 5-6 where the V_{PS} is set to 18 V and the R1 = 10 kΩ, R2 = 12 kΩ, and R3 = 15 kΩ resistors are connected in parallel.
   b. Once the circuit is activated, measure the I_{R1}, I_{R2}, and I_{R3} currents directly, as shown in the circuit of Figure 5-6(a),
   c. Measure the total output current from the power supply, $I_T = I_{R1} + I_{R2} + I_{R3}$, as shown in the circuit of Figure 5-6(b).

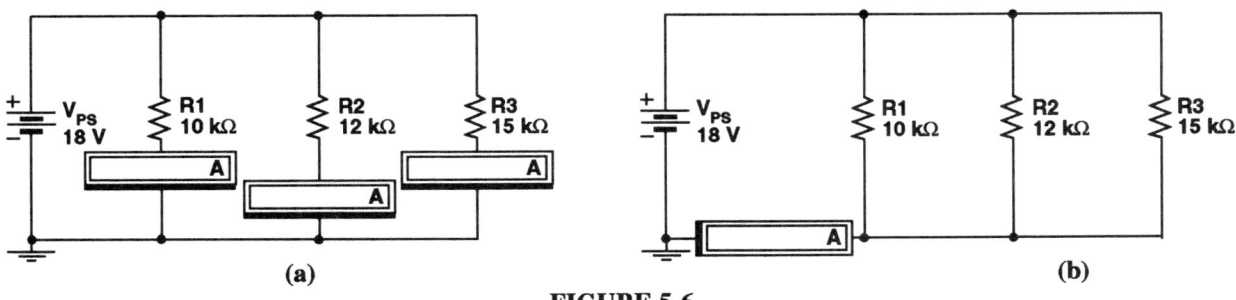

FIGURE 5-6

2. Constant Current Source:
   a. Construct, or select from the File menu, the circuit of Figure 5-7, where the constant current source is set to 4.5 mA and the R1 = 10 kΩ, R2 = 12 kΩ, and R3 = 15 kΩ resistors are connected in parallel.
   b. Once the circuit is activated, measure the I_{R1}, I_{R2}, and I_{R3} currents directly, as shown in the circuit of Figure 5-7(a),
   c. Measure the total output current, $I_T = I_{R1} + I_{R2} + I_{R3}$, as shown in Figure 5-7(b).

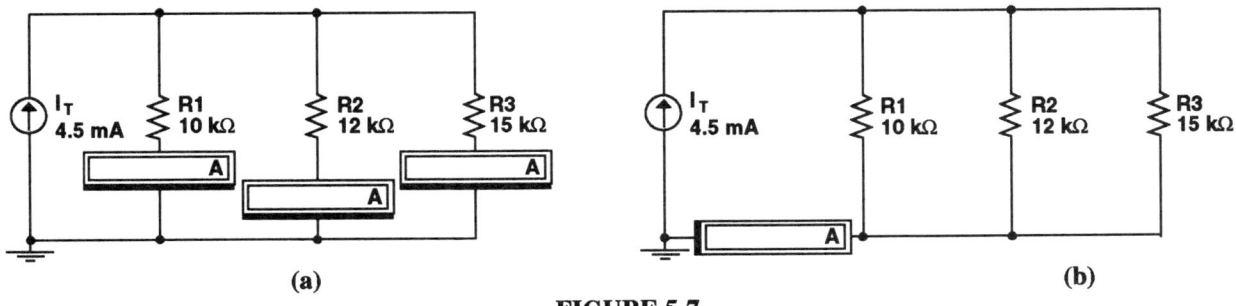

FIGURE 5-7

3. Insert the measured values, as indicated, into Table 5-2.

TABLE 5-2	Voltage Source				Current Source			
	I_{R1}	I_{R2}	I_{R3}	I_T	I_{R1}	I_{R2}	I_{R3}	I_T
CALCULATED								
MEASURED								

Questions and Problems: Basic Circuit Analysis for Electronics: 77-78

CHAPTER 6
SERIES-PARALLEL CIRCUITS

INTRODUCTION

Series-parallel circuit are solved using both series and parallel methods of analysis. For example, in the circuit connection of Figure 6-1(a) the R2 and R3 parallel resistors are solved first, allowing the circuit to be analyzed as an effective series circuit, as shown in Figure 6-1(b). Once R2 ∥ R3 and R_T are known, the total current is solved from $I_T = V_{PS}/R_T$. Then the voltage drops across R1 and R2 ∥ R3 are solved along with the currents through the individual resistors.

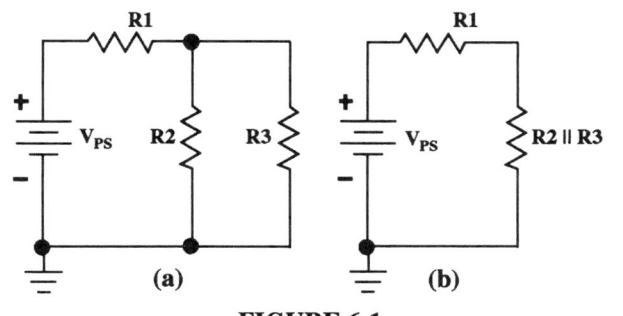

FIGURE 6-1

$R_T = R1 + (R2 \parallel R3)$
$I_T = V_{PS}/R_T$
$V_{R1} = I_T \times R1$
$V_{R2} = V_{R3} = I_T \times (R2 \parallel R3)$
$I_{R1} = V_{R1}/R1$
$I_{R2} = V_{R2}/R2$
$I_{R3} = V_{R3}/R3$
$I_T = I_{R1} = I_{R2} + I_{R3}$

Voltage divider equations can also be used to solve V_{R1} and $V_{R2} = V_{R3}$ directly, where:

$$VR1 = \frac{V_{PS} \times R1}{R1 + (R2 \parallel R3)} \qquad VR2 = VR3 = \frac{V_{PS}(R2 \parallel R3)}{R1 + (R2 \parallel R3)}$$

MORE COMPLEX CIRCUIT

The analysis of the more complex four resistor circuit is shown in conjunction with Figure 6-2. In the circuit, R2 and R3 are added and solved in parallel with R4. Then, R1 is added to solve R_T. Once R_T is known, the rest of the circuit values are solved as shown in conjunction with the circuit of Figure 6-2.

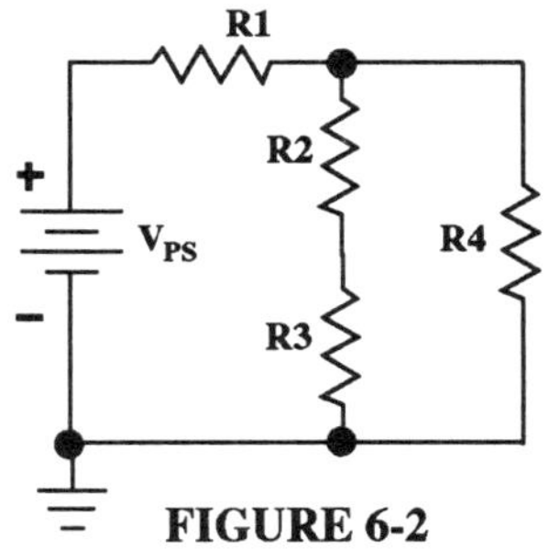

FIGURE 6-2

$R_T = R1 + [(R2 + R3) \parallel R4]$
$I_T = V_{PS}/R_T$
$V_{R1} = I_T \times R1$
$V_{(R2 + R3)} = V_{R4} = I_T \times [(R2 + R3) \parallel R4]$

$I_{R1} = V_{R1}/R1$
$I_{R2} = I_{R3} = V_{R4}/(R2 + R3)$
$I_{R4} = V_{R4}/R4$
$I_T = I_{R1} = I_{R2} + I_{R4}$

SHORTS AND OPENS

Shorts and opens cause the currents in circuits to be redistributed and the voltage drops to change. In the circuit of Figure 6-3(a), the short across R1 results in no current flow through the resistor. The open in the circuit of Figure 6-3(b) results in no current flow through the disconnected resistor R4 component.

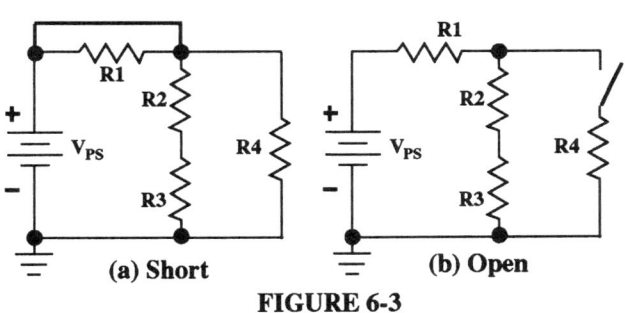

FIGURE 6-3

LABORATORY EXERCISE

READING ASSIGNMENT: Basic Circuit Analysis for Electronics: 79-84

EXERCISE OBJECTIVES
To become familiar with:

- The analysis and measurement of series-parallel circuits.
- The effects of shorts and opens on series-parallel circuits.

PROCEDURE

SECTION I (Guided Lab. Exercise): Basic Series-Parallel Circuits
PART A: Pre-laboratory calculations
1. Analyze the circuit of Figure 6-4(a). The EWB version is shown in Figure 6-4(b). In the calculations, the V_{PS} is 18 V and the resistance values are R1 at 12 kΩ, R2 at 10 kΩ, and R3 at 15 kΩ.

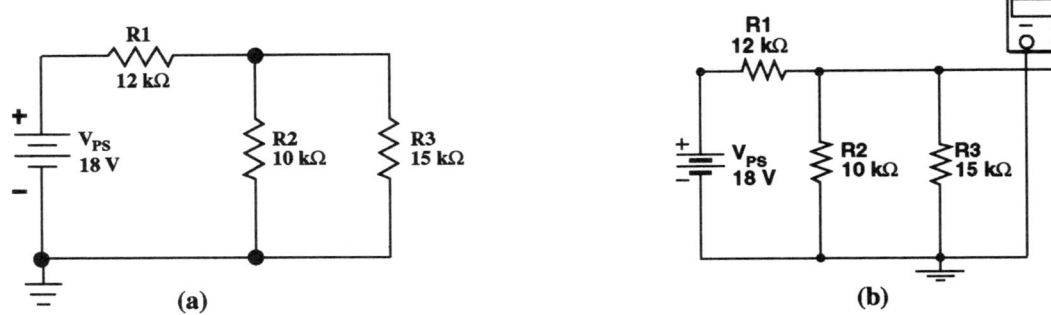

FIGURE 6-4

2. Voltage Calculations: Two methods can be used to find the voltage drops across R1, R2, and R3. In the first method Ohm's law and Kirchhoff's law are used, and in the second method Ohm's law and voltage divider equations are used. Use either method in the analysis.
   a. METHOD 1: Use Ohms law to solve R_T, I_T, and I_{R1} and I_{R2}. Then, solve V_{R1} and $V_{R2} = V_{R3}$.
      1. Calculate R_T, the total series-parallel resistance, where: $R_T = R1 + (R2 \parallel R3)$.
      2. Calculate I_T, the circuit current, where: $I_T = I_{R1} = V_{PS}/R_T$.
      3. Calculate the voltage drops across R1, where: $V_{R1} = I_T \times R1$.
      4. Calculate the voltage drop across the parallel R2 and R3 resistors where: $V_{R2} = V_{R3} = I_T(R2 \parallel R3)$.

   b. METHOD 2: Use voltage divider equation to solve V_{R1} and $V_{(R2 \parallel R3)}$ directly:

$$V_{R1} = \frac{V_{PS} \times R1}{R1 + (R2 \parallel R3)} \qquad V_{R2} = V_{R3} = \frac{V_{PS}(R2 \parallel R3)}{R1 + (R2 \parallel R3)}$$

NOTE: Voltage divider equations eliminate the need to find the current in solving V_{R1} and $V_{R2} = V_{R3}$.

3. Current Calculations: Two methods can be used to find the current through R1, R2, and R3. In the first method Ohm's law and Kirchhoff's law are used, and in the second method Ohm's law and current divider

equations are used. Use either method in the analysis.
a. METHOD 1: Calculate I_{R1}, I_{R2}, and I_{R3} where: $I_{R1} = V_{R1}/R1$, $I_{R2} = V_{R2}/R2$, and $I_{R3} = V_{R3}/R3$.
b. METHOD 2: Calculate I_{R1}, I_{R2}, and I_{R3} where: $I_{R1} = I_T$, $I_{R2} = \dfrac{I_T \times R3}{R2 + R3}$, and $I_{R3} = \dfrac{I_T \times R2}{R2 + R3}$.

NOTE: Since $I_T = I_{R1}$ is the current shared by R2 and R3, I_{R2} and I_{R3} can be solved using current divider equations:

4. Insert the calculated values R_T, V_{R1}, V_{R2}, V_{R3}, I_{R1}, I_{R2}, and I_{R3} into Table 6-1.

PART B: Voltage Measurements
1. Construct, or select from the File menu, the series-parallel circuit of Figure 6-4(b), where the V_{PS} is set to 18 V and R1 = 12 kΩ, R2 = 10 kΩ, and R3 = 15 kΩ.

2. Measure V_{R1}, V_{R2}, and V_{R3}. Make all voltage measurements with reference to ground.
   a. Measure V_{PS} at 18 V and then measure $V_{R2} = V_{R3}$.
   b. Calculate the "measured" voltage across R1 where: $V_{R1} = V_{PS} - V_{R2}$.

3. Insert the measured voltage drop values, as indicated, into Table 6-1.

PART C: Current Measurements
1. Construct, or select from the File menu, the series-parallel circuits of Figure 6-5. Then measure the currents through R1, R2, and R3 using either the direct or indirect method.

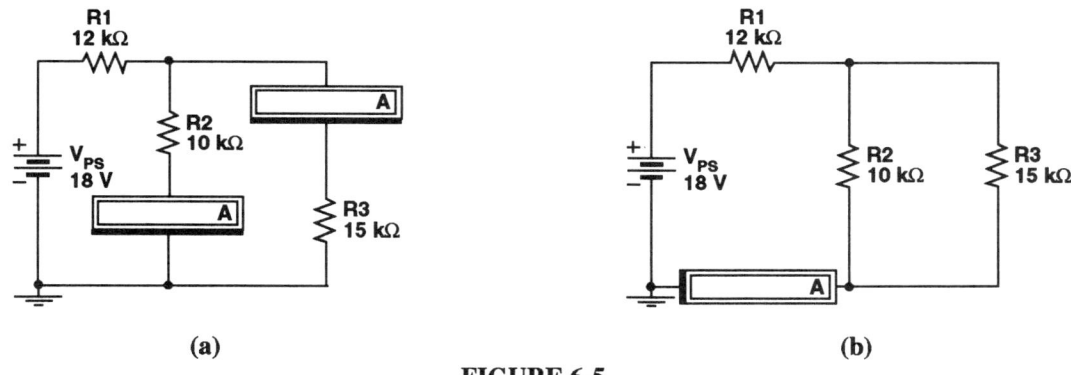

FIGURE 6-5

2. Direct and Indirect Current Measurements:
   a. Use the direct method of measuring I_{R2} and I_{R3} as shown in Figure 6-5(a). Then measure I_{R1}, which is the total current I_T through resistor R1 and from the power supply, as shown in Figure 6-5(b).
   b. In the indirect method the measured V_{R1}, V_{R2}, and V_{R3} voltages and the R1, R2, and R3 ohmic values are used to solve the I_{R1}, I_{R2}, and I_{R3} currents where: $I_{R1} = V_{R1}/R1$, $I_{R2} = V_{R2}/R2$, and $I_{R3} = V_{R3}/R3$.

3. Insert the measured values, as indicated, into Table 6-1.

TABLE 6-1	R_T	V_{R1}	V_{R2}	V_{R3}	$I_T = I_{R1}$	I_{R2}	I_{R3}
CALCULATED							
MEASURED							

40 — Basic Circuit Analysis For Electronics Using Electronics Workbench®

SECTION II: More Complex Four-Resistor Series-Parallel Circuit

PART A: Pre-laboratory Calculations

Analyze the circuit of Figure 6-6. In the calculations, the V_{PS} is 18 V and the resistances values are R1 at 2.7 kΩ, R2 at 3.3 kΩ, R3 at 4.7 kΩ, and R4 at 12 kΩ.

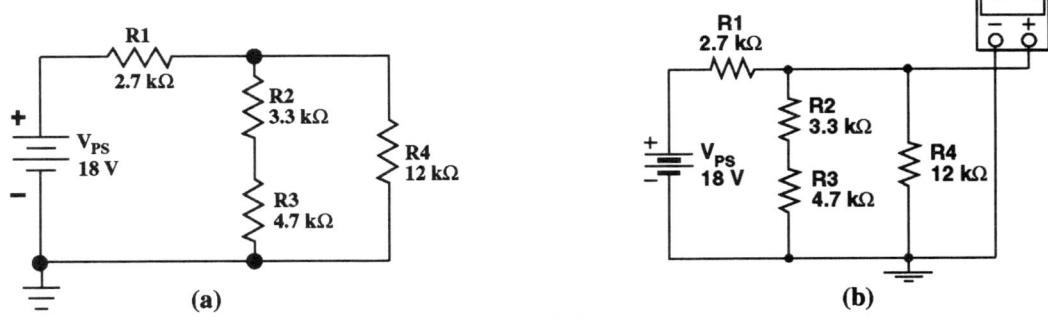

FIGURE 6-6

1. Voltage Calculations:
   a. Calculate the voltage drops across resistors R1, R2, R3, and R4 of the circuit of Figure 6-6.
   b. Use the resistance values of R1 at 2.7 kΩ, R2 at 3.3 kΩ, R3 at 4.7 kΩ, and R4 at 12 kΩ in the calculations.

2. Current Calculations:
   a. Calculate the current through resistors R1, R2, R3, and R4 of the circuit.
   b. Use the resistances values of R1 at 2.7 kΩ, R2 at 3.3 kΩ, R3 at 4.7 kΩ, and R4 at 12 kΩ in the calculations.

3. Insert the calculated voltages and current values, as indicated, into Table 6-2

PART B: Voltage Measurements

1. Construct, or select from the File menu, the series-parallel circuit of Figure 6-6(b), and measure the voltage drops of V_{R1}, V_{R2}, and V_{R3}. Make all voltage measurements with reference to ground.

2. Measure V_{PS} at 18 V, then measure V_{R3} and V_{R4}.
   a. Use the measured voltages to find the voltage across R1 where: $V_{R1} = V_{PS} - V_{R4}$.
   b. Then use the measured voltages to find the voltage across R2 where: $V_{R2} = V_{R4} - V_{R3}$.

3. Insert the measured resistor and voltage drop values, as indicated, into Table 6-2.

PART C: Current Measurements

1. Construct, or select from the File menu, the series-parallel circuit of Figure 6-7.

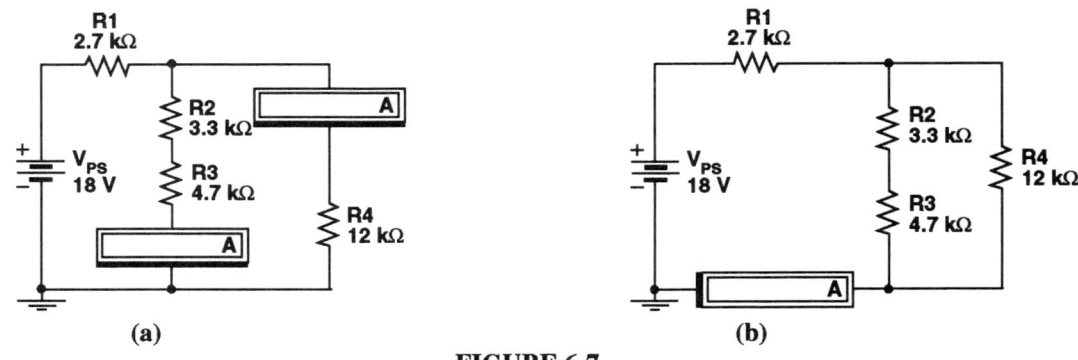

FIGURE 6-7

2. Measure the currents through R1, R2, R3, and R4 using either the direct or indirect method.

a. In the direct method measure I_{R2} and I_{R3}, as shown in Figure 6-7(a). Then measure I_{R1}, which is the total current through resistor R1 from the power supply, as shown in Figure 6-7(b).

b. In the indirect method use the measured V_{R1}, V_{R2}, and V_{R3} voltages and the R1, R2, and R3 ohmic values to solve the I_{R1}, I_{R2}, I_{R3}, and I_{R4} where: $I_{R1} = V_{R1}/R1$, $I_{R2} = I_{R3} = V_{R3}/R3$, and $I_{R4} = V_{R4}/R4$.

3. Insert the calculated and measured values, as indicated, into Table 6-2

TABLE 6-2	R_T	V_{R1}	V_{R2}	V_{R3}	$V_{(R2+R3)} = V_{R4}$	$I_T = I_{R1}$	$I_{R2} = I_{R3}$	I_{R4}
CALCULATED								
MEASURED								

Section III: Testing the Effects of Opens and Shorts
Part A: Open Circuit Conditions.

1. Analyze the circuit of Figure 6-8, where resistor R4 is open.

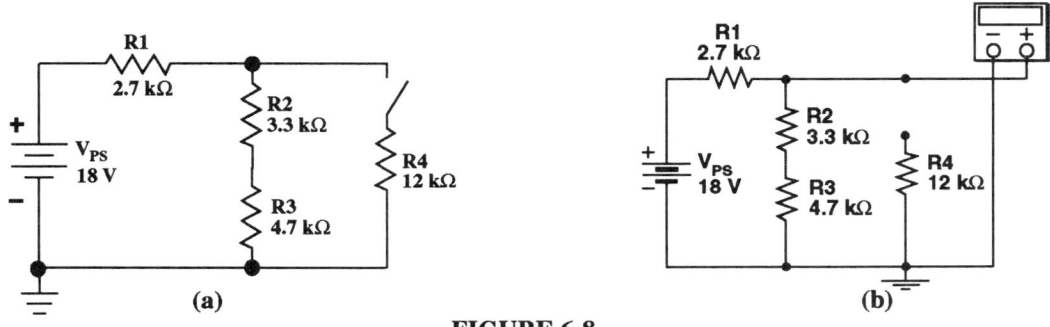

FIGURE 6-8

2. Circuit Calculations:
   a. Calculate the voltage drops across resistors R1, R2, R3, and R4 of the circuit of Figure 6-8.
   b. Calculate the current through the R1, R2, R3 and R4 resistors of the circuit.

3. Voltage Measurements:
   a. Construct, or select from the File menu, the circuit of Figure 6-8(b).
   b. Measure the voltage drops across resistors R1, R2, R3, and R4 of the circuit of Figure 6-8(b). Make all voltage measurements with reference to ground.

4. Current Measurements:
   a. Construct, or select from the File menu, the series-parallel circuit of Figure 6-9.
   b. Measure the currents through R1, R2, and R3 using the direct or indirect methods.

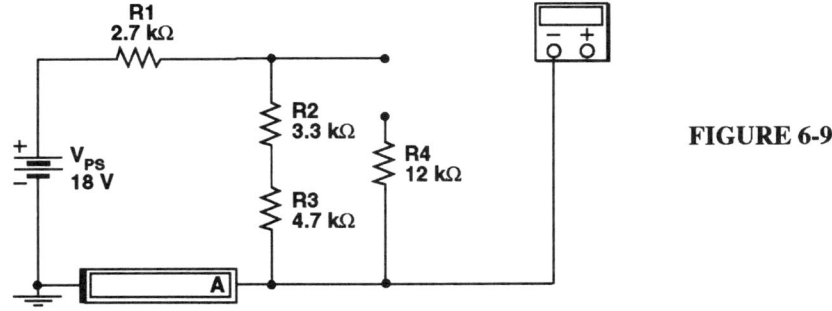

FIGURE 6-9

42 — Basic Circuit Analysis For Electronics Using Electronics Workbench®

5. Insert the calculated and measured values, as indicated, into Table 6-3.

TABLE 6-3	V_{R1}	V_{R2}	V_{R3}	V_{R4}	$I_{R1} = I_{R2} = I_{R3}$	I_{R4}
CALCULATED						
MEASURED						/////

PART B: Short Circuit Condition

1. Analyze the circuit of Figure 6-10, where a short is connected across the R1 resistor, to simulate a shorted condition that could occur while breadboarding.

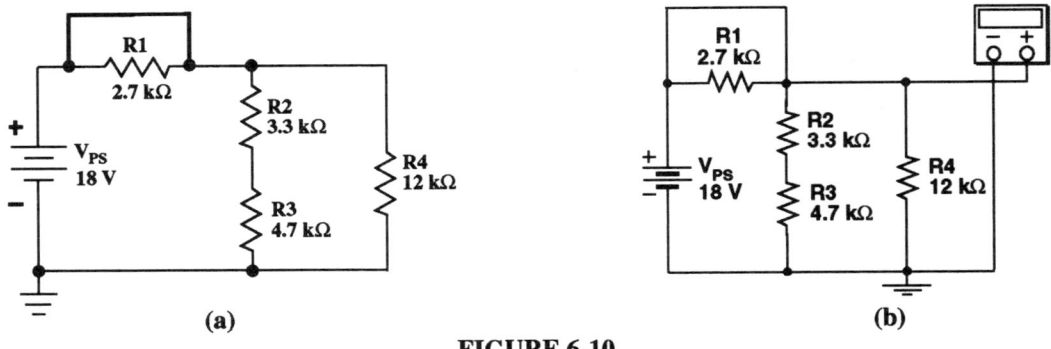

FIGURE 6-10

2. Circuit Calculations:
    a. Calculate the voltage drops across resistors R1, R2, R3 and R4 of the circuit of Figure 6-10(a).
    b. Calculate the current through the circuit resistors.

NOTE: Use the R2 = 3.3 kΩ, R3 = 4.7 kΩ, and R4 = 12 kΩ resistor values in the calculations.

3. Circuit Measurements:
    a. Construct, or select from the File menu, the circuit of Figure 6-10(b). Measure the voltage drops across resistors R1, R2, R3, and R4 of the circuit. Make all voltage measurements with reference to ground.
    b. Construct, or select from the File menu, the circuit of Figure 6-11. Measure the current through the circuit resistors. Use either the indirect or direct methods of current measurement, as shown in Figure 6-11.

4. Insert the calculated and measured values, as indicated, into Table 6-4.

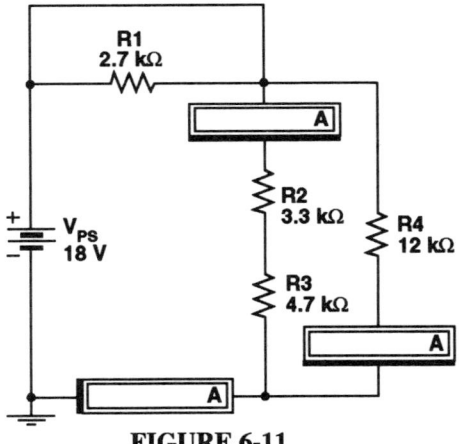

FIGURE 6-11

TABLE 6-4	V_{R1}	V_{R2}	V_{R3}	V_{R4}	I_{R1}	I_T	$I_{R2} = I_{R3}$	I_{R4}
CALCULATED								
MEASURED					/////			

SECTION IV (Optional Laboratory Exercise): Series-Parallel Circuit

1. Analyze the circuit of Figure 6-12.

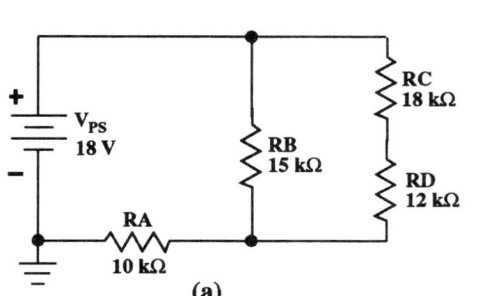
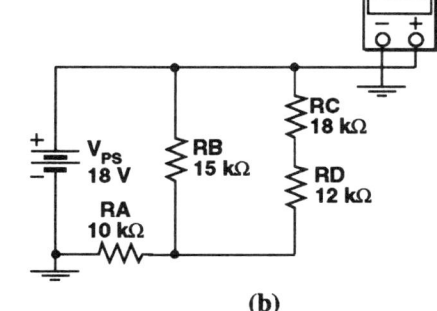

FIGURE 6-12

2. Circuit Calculations:
   a. Calculate the voltage drops across resistors RA, RB, RC, and RD of the circuit of Figure 6-12.
   b. Calculate the current through circuit resistors.

NOTE: This circuit is solved much like the previous circuits, however, no formulas are given and some research will be required.

3. Circuit Measurements:
   a. Construct, or select from the File menu, the circuit of Figure 6-12(b). Measure the voltage drops across resistors RA, RB, RC, and RD of the circuit. Make the voltage measurements with reference to ground.
   b. Construct, or select from the File menu, the circuit of Figure 6-13. Measure the current through circuit resistors. Use either the indirect or direct method of current measurement, as shown in Figure 6-13.

4. Insert the calculated and measured values, as indicated, into Table 6-5.

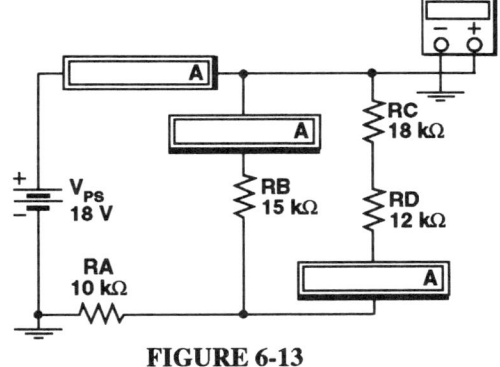

FIGURE 6-13

TABLE 6-5	R_T	V_{RA}	V_{RB}	V_{RC}	V_{RD}	$I_T = I_{RA}$	I_{RB}	I_{RC}	I_{RD}
CALCULATED									
MEASURED	/////								

Questions and Problems: Basic Circuit Analysis for Electronics: 90-91

CHAPTER 7
SERIES AIDING TWO POWER SUPPLY CIRCUITS

INTRODUCTION

Series-aiding power supplies can be used to provide a positive voltage with reference to ground as shown in Figure 7-1(a), a negative voltage with reference to ground as shown in Figure 7-1(b), or positive and negative voltages with reference to ground as shown in Figure 7-1(c).

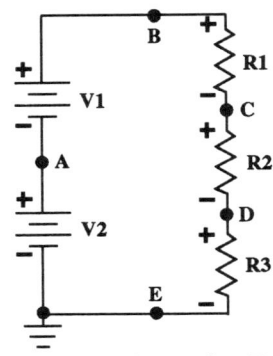
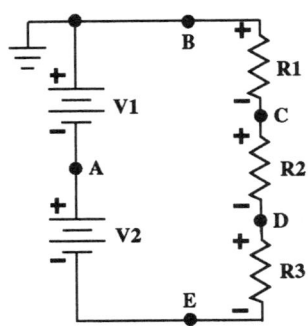
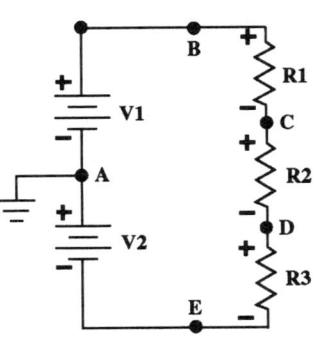

(a) Reference Ground at Point E   (b) Reference Ground at Point B   (c) Reference Ground at Point A

FIGURE 7-1

Since the only difference in the three circuits is the ground reference point, the voltage drops across the resistors are solved the same way. In the Ohm's law analysis, R_T is solved and then the voltage drops across the resistors are solved as shown below. Voltage divider equations can also be used to solve the voltage drops as shown in the second column below.

$$R_T = R1 + R2 + R3$$
$$I = (V1 + V2)/R_T$$
$$V_{R1} = I \times R1$$
$$V_{R2} = I \times R2$$
$$V_{R3} = I \times R3$$

$$V_{R1} = \frac{(V1 + V2)R1}{R1 + R2 + R3}$$

$$V_{R2} = \frac{(V1 + V2)R2}{R1 + R2 + R3}$$

$$V_{R3} = \frac{(V1 + V2)R3}{R1 + R2 + R3}$$

Ohm's Law   **Voltage Divider Equations**

So, where the ground is connected provides the positive only voltages, the negative only voltages, and the positive-and-negative voltages. The formulas used to solve the voltages with respect to reference ground for each of the connections is shown below.

$V_A = V2$	$V_A = -V1$	$V_A = 0.0\ V$
$V_B = V1 + V2$	$V_B = 0.0\ V$	$V_B = V1$
$V_C = V_B - V_{R1}$	$V_C = V_B - V_{R1}$	$V_C = V_B - V_{R1}$
$V_D = V_C - V_{R2}$	$V_D = V_C - V_{R2}$	$V_D = V_C - V_{R2}$
$V_E = 0.0\ V$	$V_E = -(V1 + V2)$	$V_E = -V2$
Reference Point E	**Reference Point B**	**Reference Point A**

LABORATORY EXERCISE

READING ASSIGNMENT: Basic Circuit Analysis for Electronics: 92-94

EXERCISE OBJECTIVES
To become familiar with:

- Two power supply circuits with reference ground at three different locations.

- Positive, negative, and positive-negative power supply connections.

PROCEDURE

SECTION I (Guided Laboratory Exercise): Positive (Series) Power Supply
PART A: Pre-laboratory Calculations
1. For the circuit of Figure 7-2(a), V1 = 12 V, V2 = 6 V, R1 = 3.3 kΩ, R2 = 4.7 kΩ, R3 = 10 kΩ, and the ground reference is at point E.

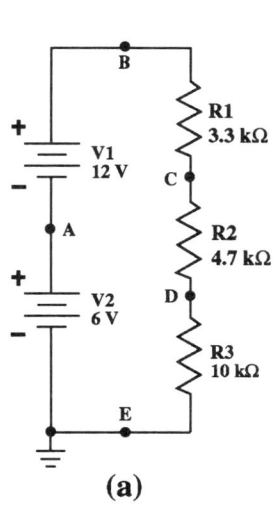
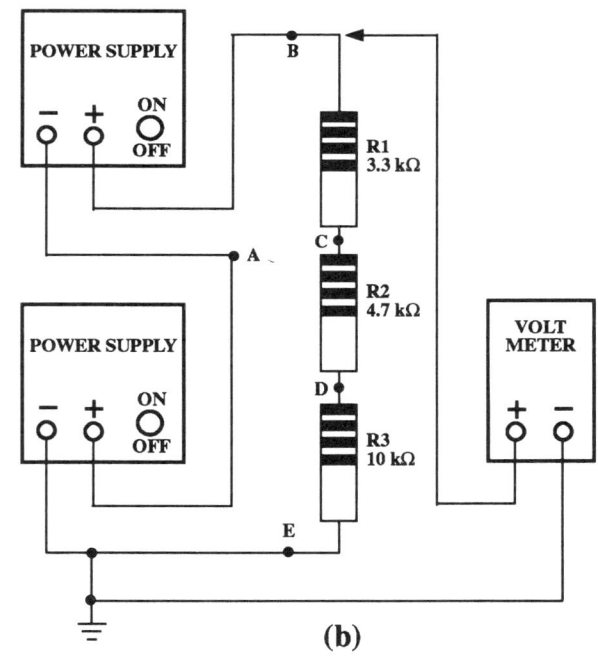

FIGURE 7-2

2. Calculate the voltage drops across the R1, R2, and R3 resistors.

   a. Calculate V_{R1}, where $V_{R1} = \dfrac{(V1 + V2)R1}{R1 + R2 + R3}$.

   b. Calculate V_{R2}, where $V_{R2} = \dfrac{(V1 + V2)R2}{R1 + R2 + R3}$.

   c. Calculate V_{R3}, where $V_{R3} = \dfrac{(V1 + V2)R3}{R1 + R2 + R3}$.

46 — Basic Circuit Analysis For Electronics Using Electronics Workbench®

3. Then, calculate the single point voltages of the circuit (V_A, V_B, V_C, and V_D), with reference to ground (point E).
   a. Calculate V_A where: $V_A = V2$.
   b. Calculate V_B where: $V_B = V1 + V2$.
   c. Calculate V_C where: $V_C = V_B - V_{R1}$, or $V_C = V_{R2} + V_{R3}$.
   d. Calculate V_D where: $V_D = V_{R3}$.

PART B: Circuit Construction and Measurements

1. Construct, or open and select from the File menu, the circuit of Figure 7-3.

2. Measure the single point voltages (V_A, V_B, V_C, and V_D) all made with respect to reference ground (point E).

3. Use the measured values of V_B, V_C, and V_D to determine the "measured" V_{R1}, V_{R2}, and V_{R3} values where: $V_{R1} = V_B - V_C$, $V_{R2} = V_C - V_D$, and $V_{R3} = V_D - V_E$.

4. Insert the calculated and measured values, as indicated, into Table 7-1.

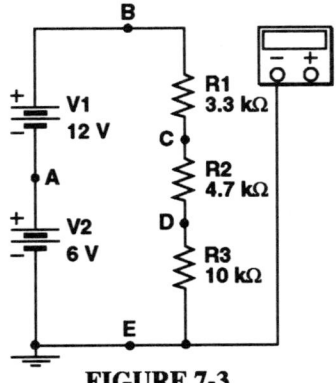

FIGURE 7-3

TABLE 7-1	V_{R1}	V_{R2}	V_{R3}	V_A	V_B	V_C	V_D
CALCULATED							
MEASURED							

SECTION II (Guided Lab. Exercise): Negative (Series) Power Supplies
PART A: Pre-laboratory Calculations

1. For the circuit of Figure 7-4(a), the circuit, the power supplies (V1 = 12 V and V2 = 6 V), and the resistor values (R1 = 3.3 kΩ, R2 = 4.7 kΩ, R3 = 10 kΩ) all remain the same, but the reference is at point B.

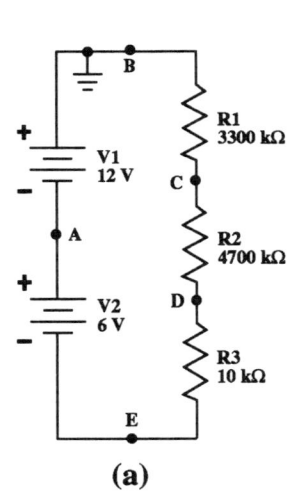

(a)

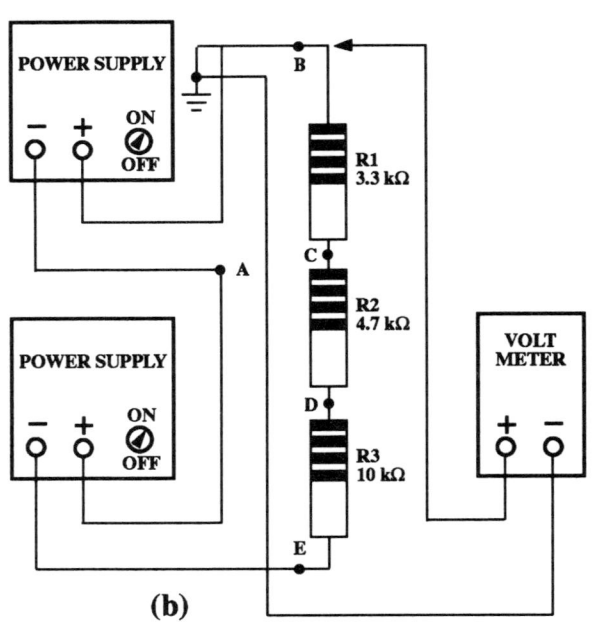

(b)

FIGURE 7-4

Series-Aiding Two Power Supply Circuits — 47

2. Since the circuit of Figure 7-4(a) is identical to the previous circuit, the voltage drops across R1, R2, and R3 are the same. Only the reference ground at point B has changed.

3. Calculate the single point voltages of the circuit (V_A, V_C, V_D and V_E) with reference to ground (point B).
   a. Calculate V_A where: $V_A = (-) V1$.
   b. Calculate V_C where: $V_C = V_B - V_{R1}$.
   c. Calculate V_D where: $V_D = V_C - V_{R2}$.
   d. Calculate V_E where: $V_E = (-)(V1 + V2)$.

PART B: Circuit Construction and Measurements
1. Construct, or open and select from the File menu, the circuit of Figure 7-5.

2. Measure the single point voltages (V_A, V_C, V_D, and V_E) all made with respect to reference ground (point B).

3. Use the measured values of V_C, V_D, and V_E to determine the "measured" V_{R1}, V_{R2}, and V_{R3} values where: $V_{R1} = V_B - V_C$, $V_{R2} = V_C - V_D$, and $V_{R3} = V_D - V_E$.

4. Insert the calculated and measured values, as indicated, into Table 7-2.

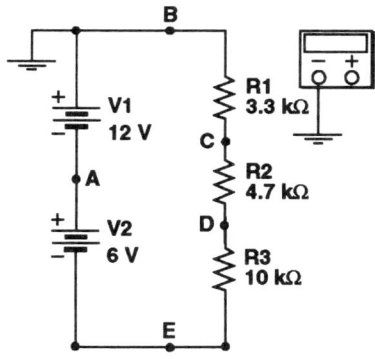

FIGURE 7-5

TABLE 7-2	V_{R1}	V_{R2}	V_{R3}	V_A	V_C	V_D	V_E
CALCULATED							
MEASURED							

SECTION III (Guided Lab. Exercise): Plus and Minus Power Supplies
PART A: Pre-Laboratory Calculations
1. For the circuit of Figure 7-6(a), the circuit, the power supply values (V1 = 12 V, V2 = 6 V), and the resistor values (R1 = 3.3 kΩ, R2 = 4.7 kΩ, R3 = 10 kΩ) all remain the same, but reference ground is at point A.

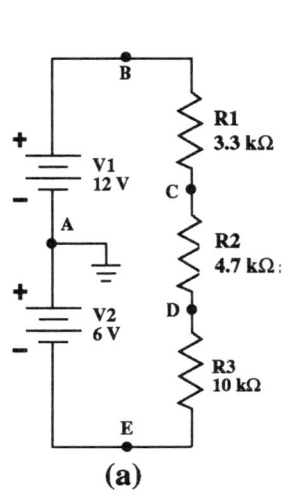

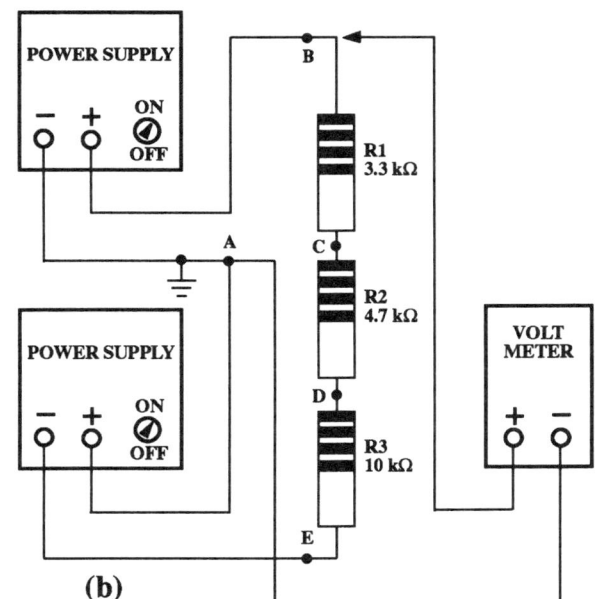

FIGURE 7-6 (a) (b)

2. Since the circuit of Figure 7-6(a) is identical to the previous circuit, the voltage drops across R1, R2, and R3 are the same. Only the reference ground at point A has changed.

3. Calculate the single point voltages of the circuit (V_B, V_C, V_D, and V_E) with reference to ground (point A).
    a. Calculate V_B, where $V_B = V1$.
    b. Calculate V_C, where $V_C = V_B - V_{R1}$.
    c. Calculate V_D, where $V_D = V_C - V_{R2}$.
    d. Calculate V_E, where $V_E = (-) V2$.

PART B: Circuit Construction and Measurements

1. Construct, or open and select from the File menu, the circuit of Figure 7-7.

2. Measure the single point voltages (V_B, V_C, V_D, and V_E) all made with respect to reference ground (point A).

3. Use the measured values of V_B, V_C, V_D, and V_E to determine the "measured" V_{R1}, V_{R2}, and V_{R3} values where: $V_{R1} = V_B - V_C$, $V_{R2} = V_C - V_D$, and $V_{R3} = V_D - V_E$.

4. Insert the calculated and measured values, as indicated, into Table 7-3.

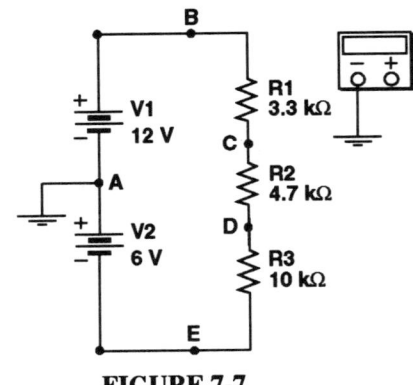

FIGURE 7-7

TABLE 7-3	V_{R1}	V_{R2}	V_{R3}	V_B	V_C	V_D	V_E
CALCULATED							
MEASURED							

SECTION IV (No Formulas Provided): Resistors Interchanged
PART A: Pre-Laboratory Calculations

1. For the plus-and-minus power supply circuit of Figure 7-8(a), the resistor values (R1 = 10 kΩ, R2 = 4.7 kΩ, R3 = 3.3 kΩ) have been changed. The power supply values (V1 = 12 V, V2 = 6 V) remain the same, as does the reference at point A.

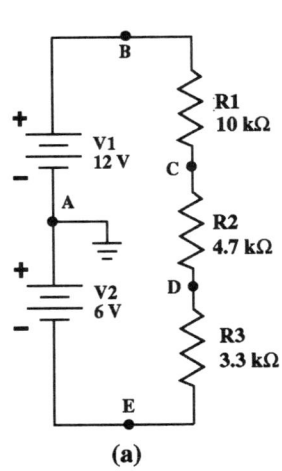

FIGURE 7-8

2. Calculate the voltage drops across the R1, R2, and R3 resistors.

3. Calculate the single point voltages of the circuit (V_B, V_C, V_D, and V_E) with reference to ground (point A).

PART B: Circuit Construction and Measurements

1. Construct, or open and select from the File menu, the circuit of Figure 7-9.

2. Measure the single point voltages (V_B, V_C, V_D, and V_E) all made with respect to reference ground (point A).

3. Use the measured values of V_B, V_C, V_D, and V_E to determine the "measured" V_{R1}, V_{R2}, and V_{R3} values.

4. Insert the calculated and measured values, as indicated, into Table 7-4.

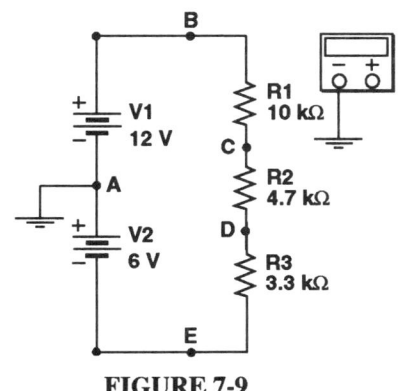

FIGURE 7-9

TABLE 7-4	V_{R1}	V_{R2}	V_{R3}	V_B	V_C	V_D	V_E
CALCULATED							
MEASURED							

SECTION V: Four Resistor Series Circuit
PART A: Pre-Laboratory Calculations

1. In the plus-and-minus power supply circuit of Figure 7-10(a) the resistor values are R1 = 12 kΩ, R2 = 15 kΩ, R3 = 10 kΩ, and R4 = 18 kΩ. Again, the power supply values (V1 = 12 V, V2 = 6 V), as well as the reference at point A, remain the same.

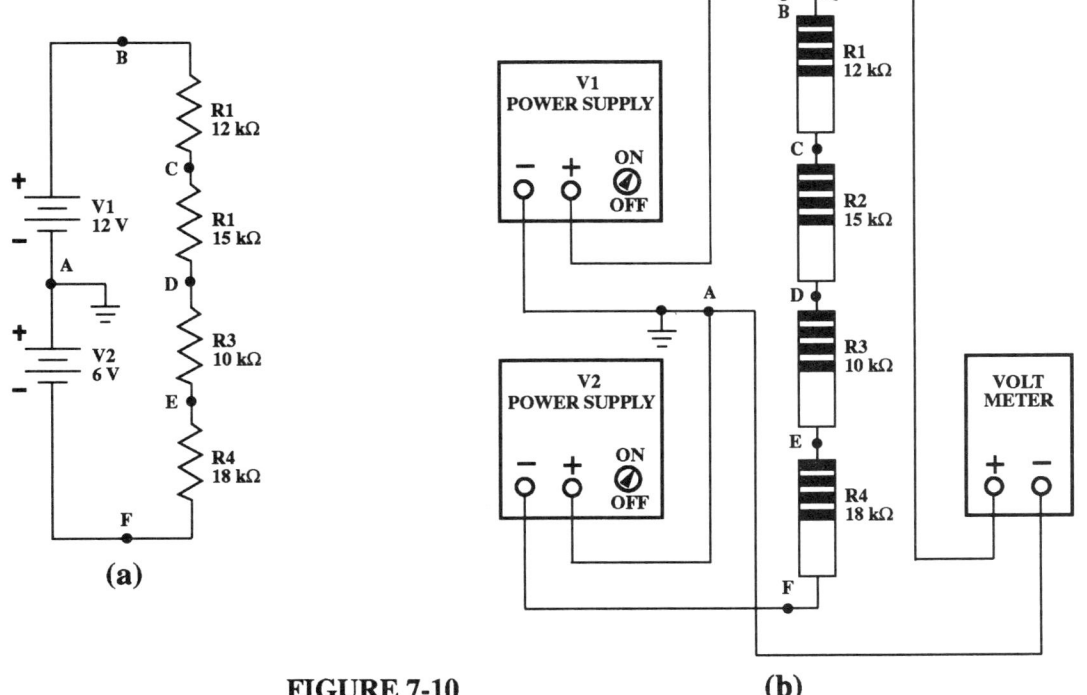

FIGURE 7-10 (a) (b)

50 — Basic Circuit Analysis For Electronics Using Electronics Workbench®

2. Calculate the voltage drops across the R1, R2, R3, and R4 resistors.

3. Calculate the single point voltages of the circuit (V_B, V_C, V_D, V_E, and V_F) with reference to ground (point A).

PART B: Circuit Construction and Measurements

1. Construct, or open and select from the File menu, the circuit of Figure 7-11.

2. Measure the single point voltages (V_B, V_C, V_D, V_E, and V_F) all made with respect to reference ground (point A).

3. Use the measured values of V_B, V_C, V_D, V_E, and V_F to determine the "measured" V_{R1}, V_{R2}, V_{R3}, and V_{R4} values.

4. Insert the calculated and measured values, as indicated, into Table 7-5.

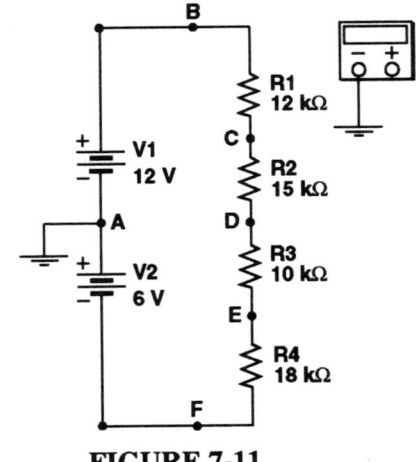

FIGURE 7-11

TABLE 7-5	V_{R1}	V_{R2}	V_{R3}	V_{R4}	V_B	V_C	V_D	V_E	V_F
CALCULATED									
MEASURED									

SECTION VI: Four Resistor Series-Parallel Circuit

PART A: Pre-laboratory Calculations

1. In the four resistor circuit of Figure 7-12(a) the resistor values are R1 = 10 kΩ in parallel with R3 = 15 kΩ, R2 = 12 kΩ, and R4 = 18 kΩ. V1 = 12 V, V2 = 6 V, and the reference at point A remains the same.

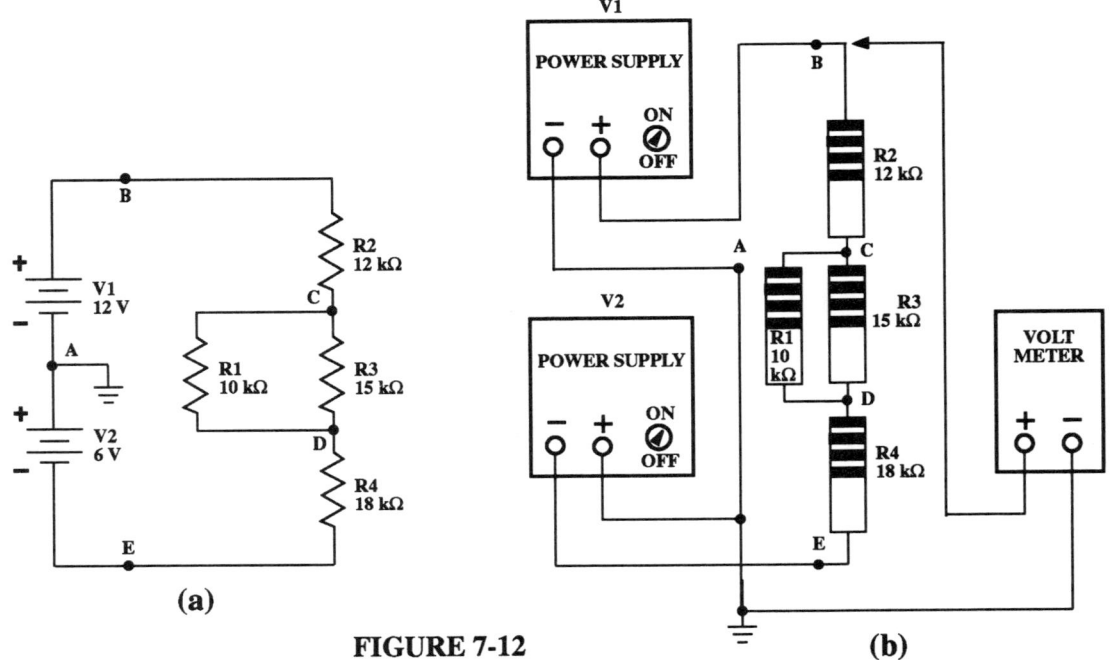

FIGURE 7-12

2. Calculate the voltage drops across the R1, R2, R3, and R4 resistors.

3. Calculate the single point voltages of the circuit (V_B, V_C, V_D, and V_E) with reference to ground (point A).

PART B: Circuit Construction and Measurements

1. Construct, or open and select from the File menu, the circuit of Figure 7-13.

2. Measure the single point voltages (V_B, V_C, V_D, and V_E). Make all measurements with respect to reference ground, which is point A, as shown in Figure 7-13.

3. Use the measured values of V_B, V_C, V_D, and V_E to determine the "measured" V_{R1}, V_{R2}, V_{R3}, and V_{R4} values.

4. Insert the calculated and measured values, as indicated, into Table 7-6.

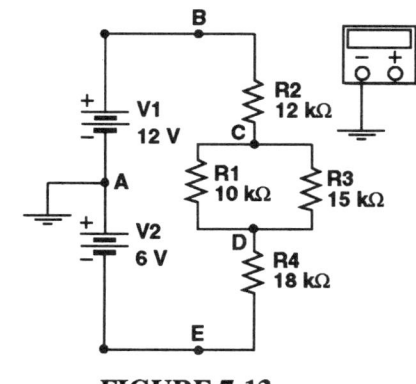

FIGURE 7-13

TABLE 7-6	V_{R1}	V_{R2}	V_{R3}	V_{R4}	V_B	V_C	V_D	V_E
CALCULATED								
MEASURED								

Questions and Problems: Basic Circuit Analysis for Electronics: 101-102

CHAPTER 8
VOLTAGE VARIABLE CIRCUITS AND APPLICATIONS

INTRODUCTION

Variable resistors are used to vary the voltage or current in circuits to provide precise voltage drops not possible with fixed resistors. Variable resistors can be connected as potentiometers or rheostats. As a Potentiometer all three terminals of the variable resistor are used, and as a rheostat two terminals are used. So, potentiometers and rheostats work differently.

By using all three terminals of the variable resistor, the potentiometer varies the output voltage of the wiper between a maximum and a minimum resistance condition. However, in the laboratory the maximum variable resistor resistance is set to 90% and the minimum to 10% of the total resistance. The formulas used to solve the voltage drops across the series resistors in the circuit and the voltage across the output at 90% and 10% variable resistance settings are shown in conjunction with the circuit of Figure 8-1(a). The two terminal rheostat is solved like the potentiometer except in the rheostat connection any change in the variable resistance causes the circuit current to change.

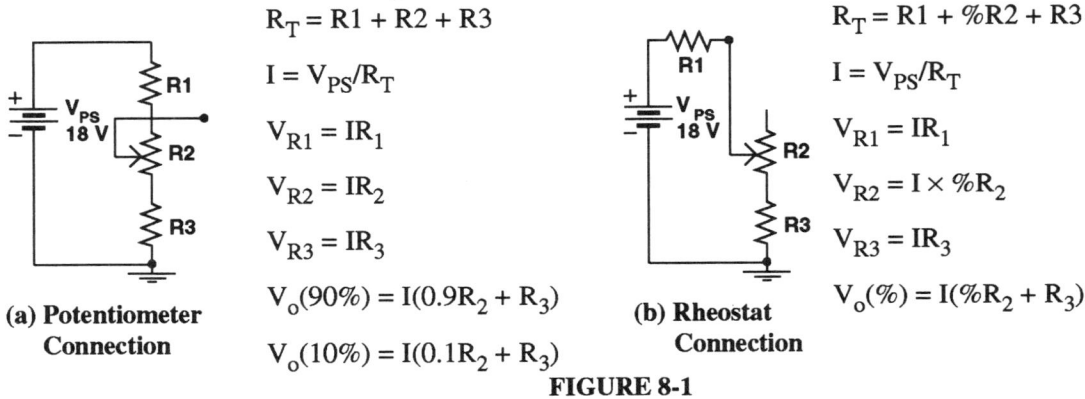

(a) Potentiometer Connection

$R_T = R1 + R2 + R3$
$I = V_{PS}/R_T$
$V_{R1} = IR_1$
$V_{R2} = IR_2$
$V_{R3} = IR_3$
$V_o(90\%) = I(0.9R_2 + R_3)$
$V_o(10\%) = I(0.1R_2 + R_3)$

(b) Rheostat Connection

$R_T = R1 + \%R2 + R3$
$I = V_{PS}/R_T$
$V_{R1} = IR_1$
$V_{R2} = I \times \%R_2$
$V_{R3} = IR_3$
$V_o(\%) = I(\%R_2 + R_3)$

FIGURE 8-1

VARIABLE RESISTORS ON THE EWB

After the variable resistor is located in the parts bin and dragged onto the workspace, the default resistance is 1 kΩ and the setting is 50%. The total resistance of the variable resistor is measured with the ohmmeter connected between terminals 1 and 3 as shown in Figure 8-2(a). The variable resistance is measured between leads 2 and 3 as shown in Figure 8-2(b). Since the default position of the variable resistance is 50%, the measured resistance will be 500 Ω. To decrease the resistance push the R key. To increase the resistance push both the shift and the R key. Another method to change the percentage is to double click on the variable resistor. When the dialog box appears, as shown in Figure 8-2(c), make the changes.

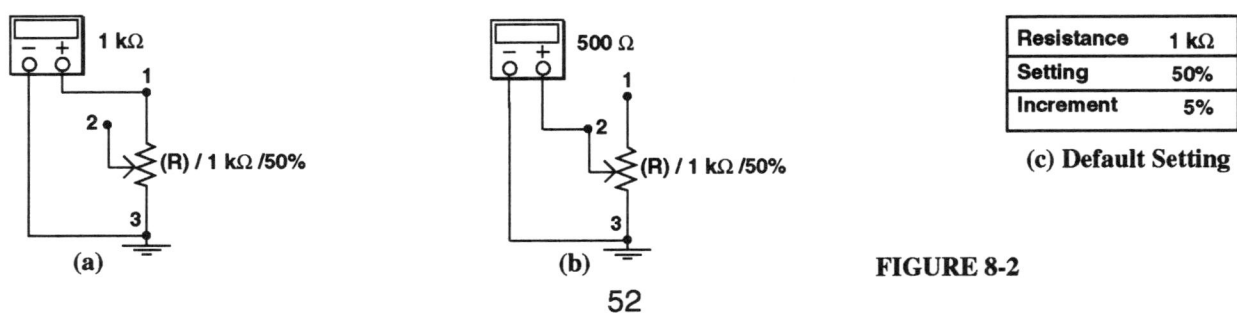

(a)   (b)   (c) Default Setting

FIGURE 8-2

52

LABORATORY EXERCISE

READING ASSIGNMENT: Basic Circuit Analysis for Electronics: 103-109

EXERCISE OBJECTIVES
To become familiar with:

- Using a variable resistor as a potentiometer to vary the output voltage.
- Using a variable resistor as a rheostat to vary the output voltage.
- The analysis and measurement of bridge circuits.

PROCEDURE

SECTION I (Guided Lab. Exercise): Voltage Variable Potentiometer
PART A : Potentiometer Set at 90% of Full CW Rotation

1. Analyze the circuit of Figure 8-3(a), where the wiper of the R2 potentiometer is set to 90% of full clockwise rotation. The EWB circuit version is shown in Figure 8-3(b). In the calculations V_{PS} is 18 V, R1 is 12 kΩ, variable resistor R2 is 10 kΩ (connected as a potentiometer at 90% of full CW rotation), and resistor R3 is 18 kΩ.

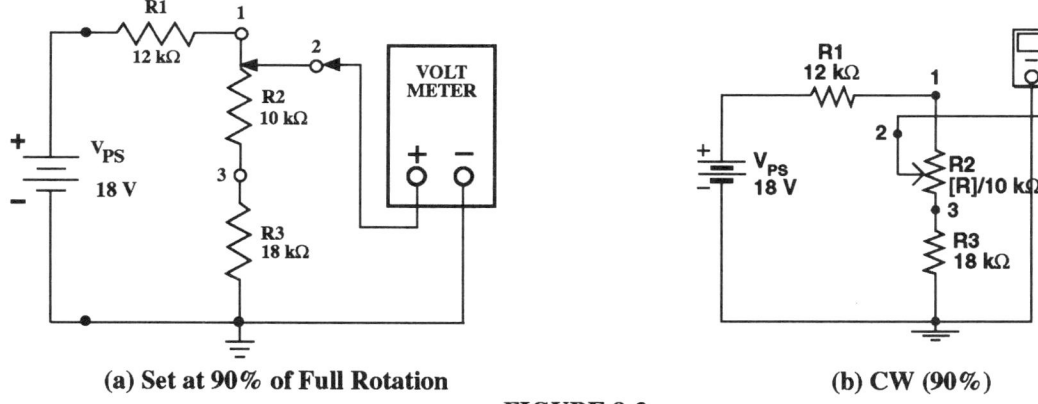

(a) Set at 90% of Full Rotation   (b) CW (90%)

FIGURE 8-3

2. Solve the voltage drops across the R1, R2, and R3 resistors. Use the standard Ohm's law equations to solve V_{R1}, V_{R2}, and V_{R3} where:

   a. $I = \dfrac{V_{PS}}{R1 + R2 + R3}$ b. $V_{R1} = IR_1$ c. $V_{R2} = IR_2$ d. $V_{R3} = IR_3$

3. Use the calculated values of V_{R1}, V_{R2}, and V_{R3} to solve the voltages at terminals 1 and 3 of the potentiometer, measured with respect to ground.
   a. Calculate V1, measured with respect to ground, from: $V1 = V_{R2} + V_{R3}$.
   b. Calculate V3, measured with respect to ground, from: $V3 = V_{R3}$.

4. Use I, the R2 resistor at 90% of full CW rotation, and R3 to solved the output voltage where:
   $V_o = I(0.9R2 + R3)$.

5. Insert the calculated values, as indicated, into Table 8-1.

54 — Basic Circuit Analysis For Electronics Using Electronics Workbench®

Voltage Measurements

1. Construct, or select from the File menu, the circuit of Figure 8-3(b), where the R2 potentiometer is set at 90% of full clockwise rotation.

2. Measure the voltage drops around the circuit. Measure V_{PS}, $V1 = V_{(R2+R3)}$, and $V3 = V_{R3}$, all measured with respect to ground. Then use the measured V_{PS}, $V_{(terminal\ 1)}$, and $V_{(terminal\ 3)}$ to determine the V_{R1}, V_{R2}, and V_{R3} voltage drops, where: $V_{R1} = V_{PS} - V1$ and $V_{R2} = V1 - V_{R3}$.

3. Connect the voltmeter to the wiper (terminal 2) of the potentiometer R2, and measure the output voltage V_o, measured with respect to reference ground.

4. Insert the measured values, as indicated, into Table 8-1.

PART B: Potentiometer Set at 10% of Full CW Rotation

1. Analyze the circuit of Figure 8-4(a), where the wiper of potentiometer R2 is set at 10% of full CW rotation. The EWB circuit version is shown in Figure 8-4(b). In the calculations V_{PS} is 18 V, R1 is 12 kΩ, variable resistor R2 is connected as a potentiometer and set to 10% of full CW rotation, and resistor R3 is 18 kΩ.

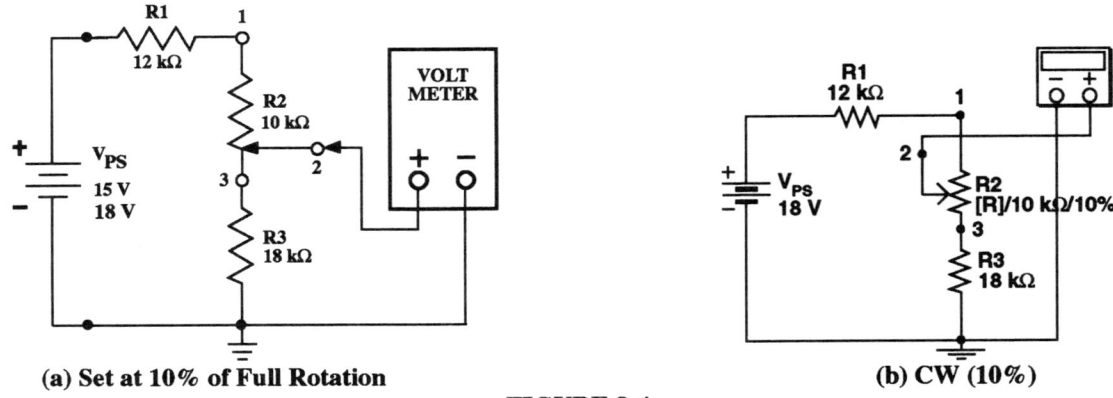

(a) Set at 10% of Full Rotation (b) CW (10%)

FIGURE 8-4

2. Use the previously solved value of current I, the R2 potentiometer set at 10% rotation, and R3 to solve the output voltage V_o, where: $V_o = I(0.1R_2 + R_3)$.

3. Insert the calculated values, as indicated, into Table 8-1.

NOTE: Potentiometers are usually constructed so, when the shaft is rotated to 90% rotation, the resistance between terminals 2 and 3 is at 90% of maximum. Then, when the shaft is rotated to 10% of full rotation, the resistance between terminals 2 and 3 is at 10% of maximum.

Voltage Measurements

1. Construct, or select from the File menu, the circuit of Figure 8-4(b), where the R2 variable resistor is at 10% of full clockwise rotation.

2. Connect the voltmeter to the wiper (terminal 2) of the R2 potentiometer, and measure the output voltage with respect to reference ground

3. Insert the measured values as indicated, into Table 8-1.

TABLE 8-1	I	V_{R1}	V_{R2}	V_{R3}	V1	V3	V_o at 90%	V_o at 10%
CALCULATED								
MEASURED								

SECTION II (Guided Laboratory Exercise): Two Terminal Rheostat
PART A: Rheostat Set at 90% of Full CW Rotation

1. Analyze the circuit of Figure 8-5(a), where the wiper of the R2 potentiometer resistance is set at 90% of the full clockwise rotation, as shown in Figure 8-5(b). In the calculations V_{PS} is 18 V, R1 is 12 kΩ, variable resistor R2 is connected as a two terminal rheostat set at 90% full rotation, and resistance R3 is 18 kΩ.

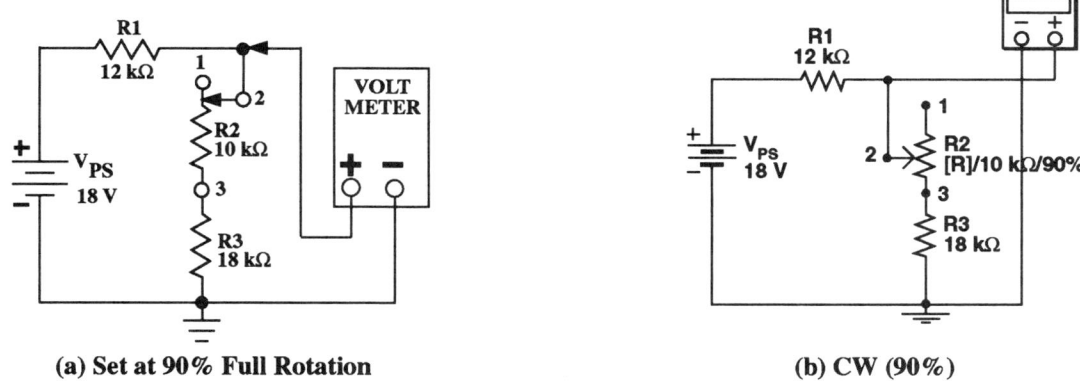

(a) Set at 90% Full Rotation (b) CW (90%)

FIGURE 8-5

2. Solve the voltage drops across R1, R2 set at 90% rotation, and R3 where:

   a. $I = \dfrac{V_{PS}}{R_1 + 0.9R_2 + R_3}$ b. $V_{R1} = IR_1$ c. $V_{R2} = I \times 0.9R_2$ d. $V_{R3} = IR_3$

3. Use the calculated values of V_{R1}, V_{R2}, and V_{R3} to solve the voltages at terminal 2 and 3 of the two terminal rheostat, measured with respect to ground.
   a. Calculate V2, measured with respect to ground, where: $V2 = V_{R2} + V_{R3}$.
   b. Calculate V3, measured with respect to ground, where: $V3 = V_{R3}$.

4. Use I, the R2 resistor at 90% of full CW rotation, and R3 to solve the V_o where: $V_o = I(0.9R_2 + R_3)$.

5. Insert the calculated values, as indicated, into Table 8-2.

NOTE: Rheostats are usually constructed so, when the shaft is rotation 90%, the resistance between terminals 2 and 3 is at 90% of maximum (for this circuit 9 kΩ).

Voltage Measurements

1. Construct, or select from the File menu, the circuit of Figure 8-5(b), where the resistance of the R2 potentiometer is set at 90% of full CW rotation.

2. Measure the voltage drops around the circuit. Measure V_{PS}, $V2 = V_{(R2 + R3)}$, and $V3 = V_{R3}$, all measured with respect to ground. Then use the measured V_{PS}, V2, and V3 to determine the V_{R1} and V_{R2} voltage drops, where: $V_{R1} = V_{PS} - V2$ and $V_{R2} = V2 - V_{R3}$.

3. Connect the voltmeter to the terminal 2 of the R2 rheostat, and measure the output voltage measured with respect to reference ground.

4. Insert the measured values, as indicated, into Table 8-2.

TABLE 8-2	I	V_{R1}	V_{R2}	V_{R3}	V2	V3	V_o at 90%
CALCULATED							
MEASURED	/////						

PART B: Rheostat Set at 10% of the Full CW Rotation

1. Analyze the circuit of Figure 8-6(a), where wiper of the R2 rheostat is set at 10% of full clockwise rotation. The EWB circuit version is shown in Figure 8-6(b). In the calculations V_{PS} is 18 V, R1 is 12 kΩ, variable resistor R2 is connected as a rheostat set at 10% of full CW rotation, and the resistance of R3 is 18 kΩ.

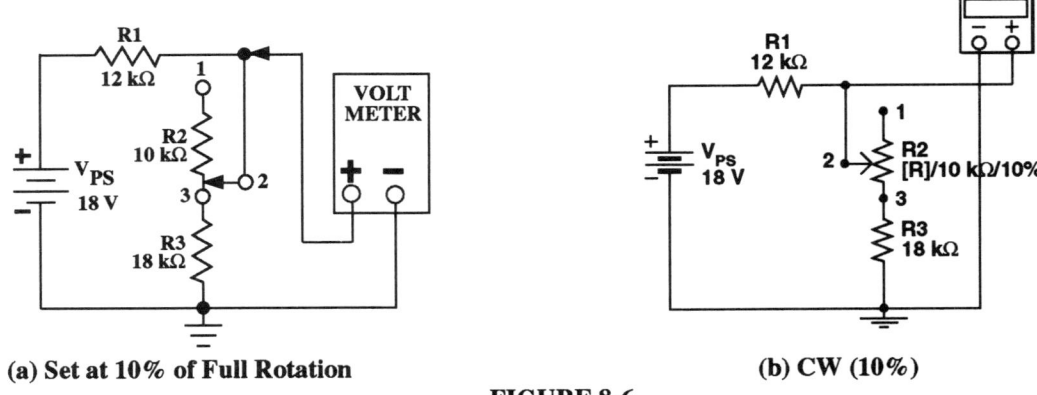

(a) Set at 10% of Full Rotation   (b) CW (10%)

FIGURE 8-6

2. Solve the voltage drops across R1, R2 set at 10% full CW rotation, and R3. Use Ohm's law equations to solve V_{R1}, V_{R2}, and V_{R3}, where:

    a. $I = \dfrac{V_{PS}}{R_1 + 0.1R_2 + R_3}$ b. $V_{R1} = IR_1$ c. $V_{R2} = I \times 0.1R_2$ d. $V_{R3} = IR_3$

3. Use the calculated values of V_{R1}, V_{R2}, and V_{R3} to solve the voltages at terminals 2 and 3 of the two terminal rheostat, measured with respect to ground.
   a. Calculate V2, measured with respect to ground, where: $V2 = V_{R2} + V_{R3}$.
   b. Calculate V3, measured with respect to ground, where: $V3 = V_{R3}$.

4. Use I, the R2 resistor set at 10% full clockwise rotation, and R3 to solve the output voltage, where:
   $V_o = I(0.1R_2 + R_3)$

5. Insert the calculated values, as indicated, into Table 8-3.

Voltage Measurements

1. Construct, or select from the File menu, the circuit of Figure 8-6(b), where the R2 variable resistor is connected as a rheostat at 10% of full clockwise rotation.

2. Connect the voltmeter to terminal 2 of the R2 rheostat, and measure the output voltage V_o, measured with respect to ground.

3. Insert the measured values, as indicated, into Table 8-3.

NOTE: Rheostats are usually constructed so, when the shaft is rotated 10% of full rotation, the resistance between terminals 2 and 3 is at 10% of maximum (for this circuit 1 kΩ).

TABLE 8-3	I	V_{R1}	V_{R2}	V_{R3}	V2	V3	V_o at 10%
CALCULATED							
MEASURED							

SECTION III: (Guided Lab. Exercise): Unbalanced Bridge Circuits

PART A: Pre-laboratory Calculations

1. Analyze the circuit of Figure 8-7(a). The EWB version is shown in Figure 8-7(b). In the calculations V_{PS} is 18 V, R1 is 2.7 kΩ, R2 is 3.3 kΩ, R3 is 18 kΩ, and R4 is 12 kΩ.

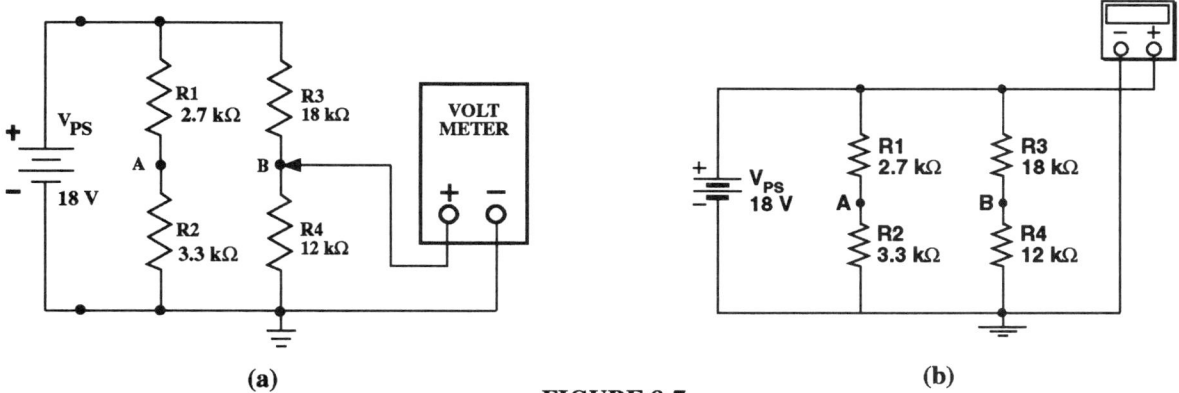

FIGURE 8-7

2. Solve the voltage drops across R1 and R2. Use voltage divider equations to solve V_{R1} and $V_{R2} = V_A$ where:

    a. $V_{R1} = \dfrac{V_{PS} \times R1}{R1 + R2}$ $\qquad V_{R2} = \dfrac{V_{PS} \times R2}{R1 + R2}$

3. Solve the voltage drops across R3 and R4. Use voltage divider equations to solve V_{R3} and $V_{R4} = V_B$ where:

    a. $V_{R3} = \dfrac{V_{PS} \times R3}{R3 + R4}$ $\qquad V_{R4} = \dfrac{V_{PS} \times R4}{R3 + R4}$

4. Calculate the output voltage V_o, the difference voltage between the V_A and V_B terminals.

5. Calculate $I_{R1} = I_{R2}$, I_{R3}, and I_{R4} where: $I_{R1} = I_{R2} = V_{R2}/R2$ and $I_{R3} = I_{R4} = V_{R4}/R4$.

6. Insert the calculated values, as indicated, into Table 8-4.

PART B. Voltage Measurements

1. Construct, or select from the File menu, the circuit of Figure 8-7(b), and measure the voltage drops around the circuit.
    a. Measure V_{PS}, $V_A = V_{R2}$, and $V_B = V_{R4}$, measured with respect to ground.
    b. Use the measured V_A and V_B voltages to determine the "measured" voltage drops of V_{R1} and V_{R3} where: $V_{R1} = V_{PS} - V_A$ and $V_{R2} = V_{PS} - V_B$.

2. Measure V_o the voltage between V_A and V_B indirectly, using the measured V_A and V_B and determine V_o, where: $V_o = V_A - V_B$.

3. Insert the measured values, as indicated, into Table 8-4.

NOTE: Although V_o can be measured directly across terminals A and B, with a floating (not grounded) voltmeter, it promotes good measurement practice to make all measurements with respect to ground.

TABLE 8-4	V_{R1}	$V_A = V_{R2}$	V_{R3}	$V_B = V_{R4}$	$I_{R1} = I_{R2}$	$I_{R3} = I_{R4}$	V_o
CALCULATED							
MEASURED							

58 — Basic Circuit Analysis For Electronics Using Electronics Workbench®
SECTION IV (Guided Laboratory Exercise): Balanced Bridge Circuits
PART A: Pre-Laboratory Calculations
1. Analyze the balanced bridge circuit of Figure 8-8(a). The EWB circuit version is shown in Figure 8-8(b). In the calculations use a V_{PS} of 18 V, R1 is 2.7 kΩ, R2 is 10 kΩ (rheostat used as a variable resistor), R3 is 18 kΩ, and R4 is 12 kΩ.

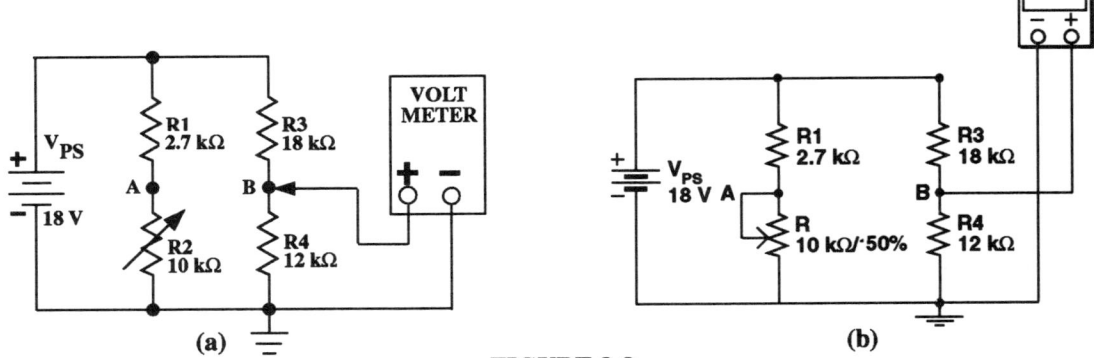

FIGURE 8-8

2. Solve the resistance value of the R2 resistance to achieve balanced condition, where $R2 = \dfrac{R1 \times R4}{R3}$.

3. Once R2 is known, verify balanced conditions. Use the voltage divider equations to solve V_{R2}, where:
$$V_A = V_{R2} = \dfrac{V_{PS} \times R2}{R1 + R2}$$

4. Solve the voltage drops across R4. Use the voltage divider equations to solve $V_{R4} = V_B$ where:
$$V_B = V_{R4} = \dfrac{V_{PS} \times R4}{R3 + R4}$$

5. Calculate the output voltage V_o at balanced conditions of $V_o = V_A - V_B$ where: $V_A = V_{R2}$ and $V_B = V_{R4}$.

6. Insert the calculated values, as indicated, into Table 8-5.

PART B. Voltage Measurements
1. Construct, or open and select from the File menu, the circuit of Figure 8-8(b), and measure the voltage drops around the circuit.
   a. Measure V_{PS} and $V_B = V_{R4}$, measured with respect to ground
   b. Then measure V_A while varying the variable R2 resistor resistance. Adjust V_A so $V_A = V_B$.

2. Disconnect the variable resistor from the balanced bridge circuit ($V_o = V_A - V_B = 0.0$ V), and measure the variable resistance.

3. Insert the measured values, as indicated, into Table 8-5.

TABLE 8-5	R2	$V_A = V_{R2}$	$V_B = V_{R4}$	V_o
CALCULATED				
MEASURED				

Questions and Problems: Basic Circuit Analysis for Electronics: 115-116

CHAPTER 9
THEVENIN'S AND NORTON'S EQUIVALENT CIRCUITS AND MAXIMUM POWER TRANSFER

INTRODUCTION

Thevenin's and Norton's equivalent circuits are widely used in the analysis of complex circuits. Thevenin's equivalent reduces any complex circuit to a voltage source (V_{TH}), a single series resistance (R_{TH}), and a load resistor (R_L). Norton's equivalent reduces a complex circuit to a constant current source, a single parallel resistance (R_N), and the load resistor. Maximum power transfer for the Thevenin equivalent circuit occurs when $R_{TH} = R_L$. For the Norton equivalent circuit, maximum power transfer occurs when $R_N = R_L$.

THEVENIN'S EQUIVALENT CIRCUIT

The series-parallel circuit of Figure 9-1 is solved using the Thevenin techniques. First, the load resistor is removed ($R_L = R3$) and the voltage across the output terminals is solved ($V_{TH} = V_{R2}$). Then, the series resistor R_{TH} is solved from the parallel combination of R1 and R2. Finally, $V_{RL} = V_{R3}$ is solved using voltage divider equations. The calculations are shown in conjunction with the circuit of Figure 9-1.

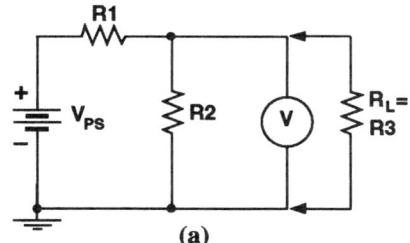 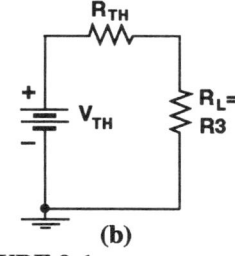

$$V_{TH} = V_{R2} = \frac{V_{PS} \times R2}{R1 + R2}$$

$$R_{TH} = R1 \parallel R2$$

$$V_{RL} = V_{R3} = \frac{V_{TH} \times R_L}{R_{TH} + R_L}$$

FIGURE 9-1

NORTON'S EQUIVALENT CIRCUIT

The series-parallel circuit of Figure 9-2 is solved using Norton techniques. First, the load is removed ($R_L = R3$) and the output current is solved by placing an ammeter, an effective short, across the output terminals so $I_N = V_{PS}/R1$. Then, the parallel resistor R_N is solved using the parallel combination of R1 and R2, I_{RL} is solved using current divider equations, and V_{RL} is solved using Ohm's law. The equations are shown in conjunction with Figure 9-2.

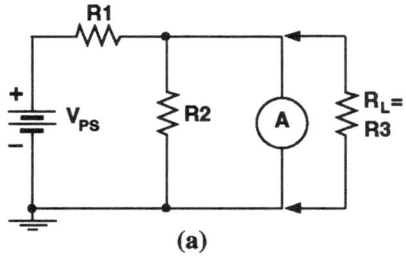 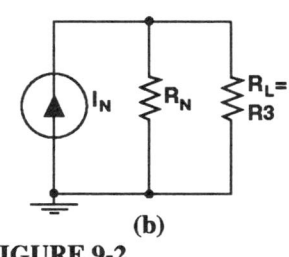

$$I_N = \frac{V_{PS}}{R1}$$

$$R_N = R1 \parallel R2$$

$$I_{RL} = \frac{I_N \times R_N}{R_N + R_L}$$

$$V_{RL} = I_{RL} \times R_L$$

FIGURE 9-2

60 — Basic Circuit Analysis For Electronics Using Electronics Workbench®

LABORATORY EXERCISE

READING ASSIGNMENT: Basic Circuit Analysis for Electronics: 117-126

EXERCISE OBJECTIVES
To become familiar with:

- Using Thevenin's theorem to reduce multi-resistor circuits to an equivalent circuit, containing a single voltage source, a source resistance, and a load resistor.

- Proving the maximum power transfer theorem and plotting a power response curve.

PROCEDURE

Section I (Guided Laboratory Exercise): Series-Parallel Circuit
PART A: Standard Analysis Techniques
1. Analyze the series-parallel circuit of Figure 9-3(a). The EWB version is shown in Figure 9-3(b). In the calculations, the V_{PS} is 18 V and circuit resistances are R1 at 27 kΩ, R2 at 33 kΩ, and R3 at 47 kΩ.

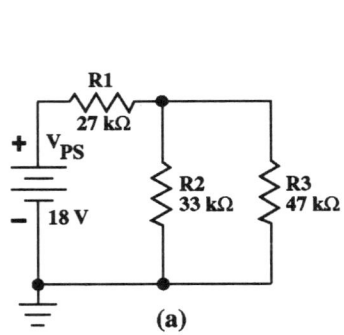 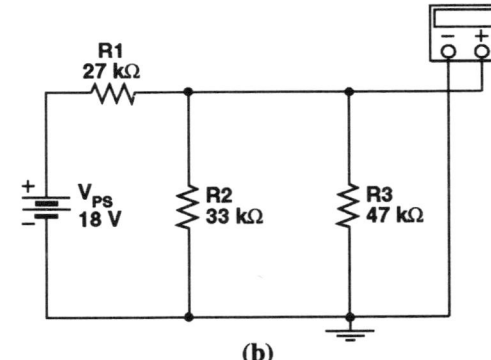

FIGURE 9-3

2. Calculate V_{R1}, V_{R2}, and V_{R3}.

   a. Calculate V_{R1} where: $V_{R1} = \dfrac{V_{PS} \times R1}{R1 + (R2 \parallel R3)}$.

   b. Calculate $V_{R2} = V_{R3}$ where: $V_{R2} = V_{R3} = \dfrac{V_{PS}(R2 \parallel R3)}{R1 + (R2 \parallel R3)}$.

3. Insert the calculated circuit voltages and currents into Table 9-1.

PART B: Voltage Measurements
1. Connect, or open and select from the File menu, the circuit of Figure 9-3(b).

2. Measure V_{PS} and $V_{R2} = V_{R3}$. Then determine the "measured" V_{R1} voltage where: $V_{R1} = V_{PS} - V_{R2}$.

3. Insert the measured values, as indicated, into Table 9-1.

PART C (Guided Pre-laboratory Calculations): Thevenin's Equivalent Circuit Analysis
1. Calculate the circuit voltages of Figure 9-4, where the R3 "load" resistor has been removed.

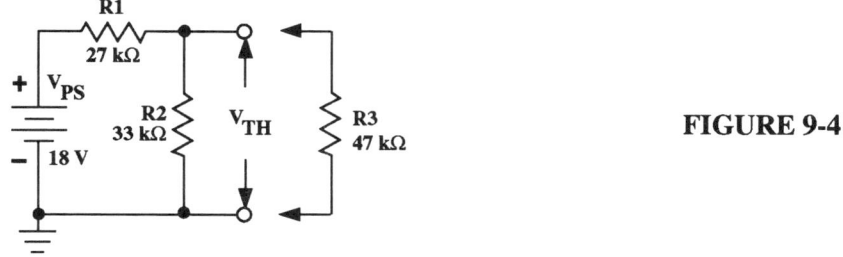

FIGURE 9-4

2. Calculate V_{TH}, R_{TH}, and V_{R_L} for the equivalent circuit of Figure 9-4.
   a. Calculate V_{TH}, the voltage drop across R2, once the load resistor R3 has been removed from the circuit.

   Calculate where: $V_{TH} = V_{R2} = \dfrac{V_{PS} \times R2}{R1 + R2}$

   b. Calculate R_{TH}, the theoretical parallel combination of R1 and R2, where $R_{TH} = R1 \parallel R2$.
   c. Calculate V_{R_L}, the voltage across the Thevenin's equivalent circuit load, where: $V_{R_L} = \dfrac{V_{TH} \times R_L}{R_{TH} + R_L}$.

3. Insert the calculated values, as indicated, into Table 9-1.

PART D: Measure R_{TH} Indirectly
1. Construct, or open and select from the File menu, the circuit of Figure 9-5.

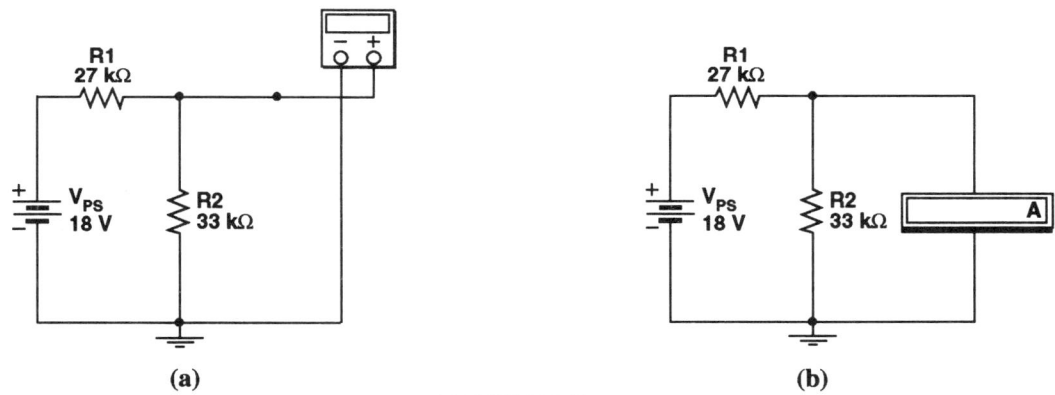

FIGURE 9-5

2. Measure V_{TH} and I_{SC} in solving R_{TH}.
   a. Measure V_{TH}, the voltage drop across the output terminals and across R2, as shown in Figure 9-5(a).
   b. Measure $I_{R1} = I_{SC}$ by connecting an ammeter across the output terminals, as shown in Figure 9-5(b).
   c. Use the measured V_{TH} and $I_{R1} = I_{SC}$ values to determine R_{TH} where: $R_{TH} = V_{TH}/I_{R1}$.

NOTE: Placing an ammeter across the output terminals and R2, effectively shorts out R2, and places the ammeter in series with the R1 resistor. Therefore the I_{SC} current flow through the ammeter is limited to I_{R1}.

3. Insert the measured values, as indicated, into Table 9-1.

PART E: Verify V_{R1} Using the Equivalent Circuit
1. Construct, or open and select from the File menu, the circuit of Figure 9-6(b).

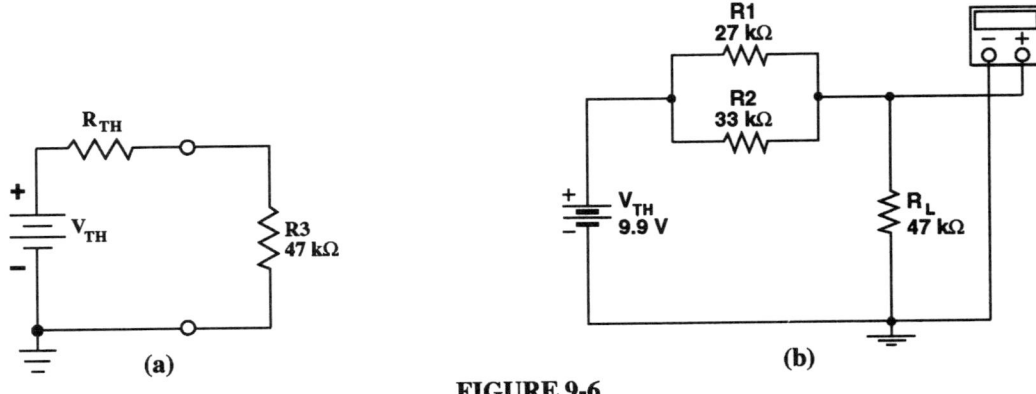

FIGURE 9-6

2. In the equivalent circuit the power supply is set to the V_{TH} voltage, the R_{TH} resistance is (R1 ∥ R2), and the load resistance is R_L = R3 at 47 kΩ.

3. Measure V_{R_L}, the voltage across the R3 load resistor of the equivalent circuit.

4. Insert the measured values, as indicated, into Table 9-1.

TABLE 9-1	Standard Analysis FIGURE 9-3		Thevenin Equivalent Circuit Analysis FIGURE 9-4 and FIGURE 9-5			Equivalent FIGURE 9-6
	V_{R1}	$V_{R2} = V_{R3}$	V_{TH}	I_{SC}	R_{TH}	V_{R_L}
CALCULATED				/////		
MEASURED						

SECTION II: More Complex Series-Parallel Circuit Analysis
PART A: Standard Analysis Techniques

1. Analyze the series-parallel circuit of Figure 9-7(a). The EWB version is shown in Figure 9-7(b). In the calculations, the V_{PS} is 18 V and circuit resistors are R1 at 27 kΩ, R2 at 33 kΩ, R3 at 12 kΩ, and R4 at 15 kΩ.

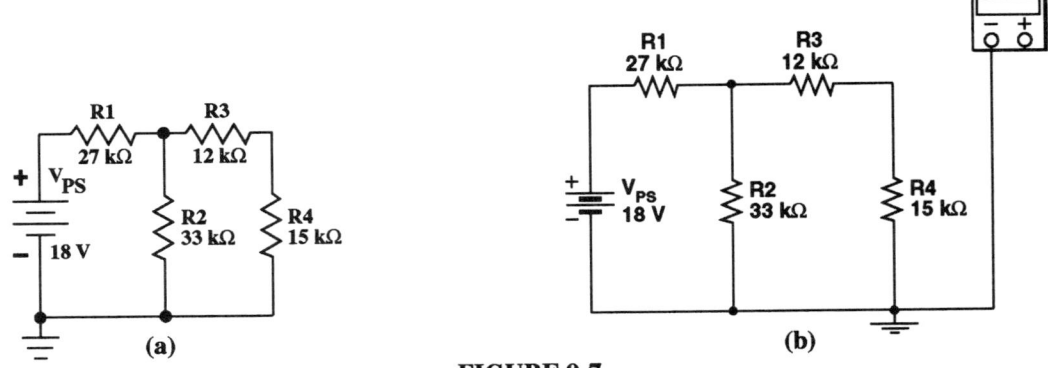

FIGURE 9-7

2. Calculate V_{R1}, V_{R2}, V_{R3}, and V_{R4} using voltage divider equations.

   a. Calculate V_{R1} where: $V_{R1} = \dfrac{V_{PS} \times R1}{R1 + (R2 \parallel [R3 + R4])}$.

   b. Calculate $V_{R2} = \dfrac{V_{PS} \times (R2 \parallel [R3 + R4])}{R1 + (R2 \parallel [R3 + R4])}$.

Thevenin's and Norton' Equivalent Circuits and Maximum Power Transfer — 63

c. Calculate $V_{R3} = \dfrac{V_{R2} \times R3}{R3 + R4}$.

d. Calculate $V_{R4} = \dfrac{V_{R2} \times R4}{R3 + R4}$.

3. Insert the calculated circuit voltages, as indicated, into Table 9-2.

PART B: Voltage Measurements

1. Connect, or open and select from the File menu, the circuit of Figure 9-7(b).

2. Measure V_{PS} and V_{R2} and V_{R4}.
   a. Determine the "measured" circuit voltage of V_{R1} where: $V_{R1} = V_{PS} - V_{R2}$.
   b. Determine the "measured" circuit voltage of V_{R3} where: $V_{R3} = V_{R2} - V_{R4}$.

3. Insert the measured values, as indicated, into Table 9-2.

PART C (Guided Pre-laboratory Calculations): Thevenin's Equivalent Circuit Analysis

1. Calculate the circuit voltages of Figure 9-8, with the R4 "load" resistor removed.

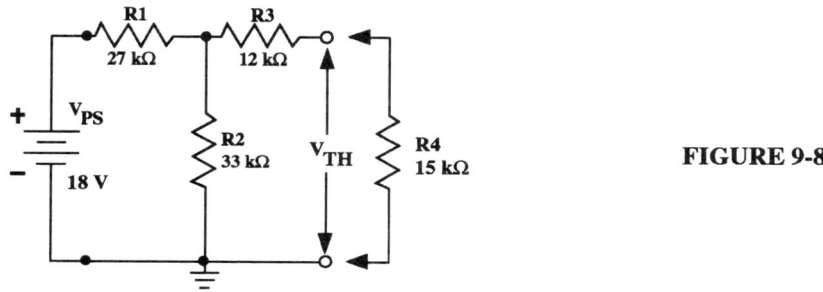

FIGURE 9-8

2. Calculate V_{TH}, R_{TH}, and V_{R_L} for the equivalent circuit of Figure 9-8.
   a. Calculate V_{TH}, the voltage across the output, which is effectively V_{R2} once the load resistor R4 has been removed where:
   $$V_{TH} = V_{R2} = \dfrac{V_{PS} \times R2}{R1 + R2}.$$

   b. Calculate R_{TH}, the theoretical combination of resistor R3 in series with resistors R1 and R2 in parallel, where: $R_{TH} = R3 + (R1 \| R2)$.

   c. Calculate V_{R_L}, the voltage across the Thevenin's equivalent circuit load, where:
   $$V_{R_L} = \dfrac{V_{TH} \times R_L}{R_{TH} + R_L}.$$

3. Insert the calculated values, as indicated, into Table 9-2.

PART D: Determine R_{TH} Indirectly Using the Measured V_{TH} and I_{SC} Values

1. Measure V_{TH}, the voltage drop across the output terminals, as shown in Figure 9-9(a).

2. Measure $I_{R3} = I_{SC}$ by connecting an ammeter across the output terminals, as shown in Figure 9-9(b).

3. Use the measured V_{TH} and $I_{R3} = I_{SC}$ values to determine R_{TH} where: $R_{TH} = V_{TH}/I_{R3}$.

64 — Basic Circuit Analysis For Electronics Using Electronics Workbench®

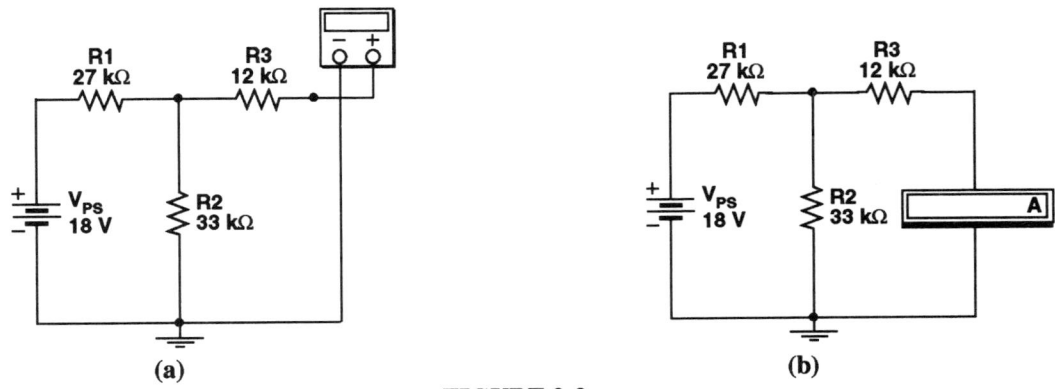

FIGURE 9-9

4. Insert the measured values, as indicated, into Table 9-2.

PART E: Verify V_{RL} using the Equivalent Circuit

1. Construct, or open and select from the File menu, the circuit of Figure 9-10.

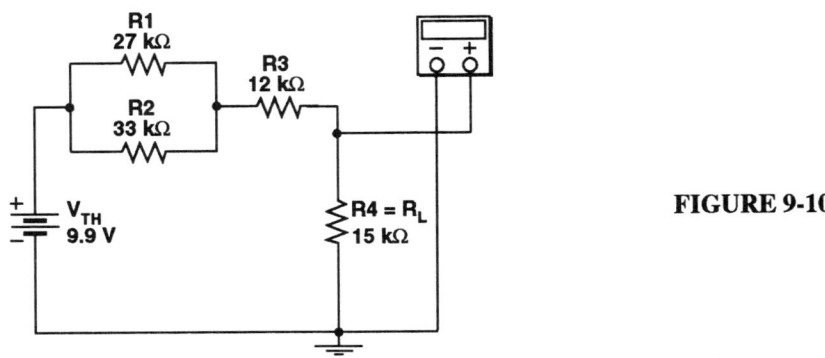

FIGURE 9-10

2. In the equivalent circuit the power supply is set to the V_{TH} voltage, the R_{TH} resistance is R3 + (R1 ∥ R2), and the load resistance R_L = R4 at 15 kΩ.
3. Measure V_{RL}, the voltage across the load resistor of the equivalent circuit
4. Insert the measured values, as indicated, into Table 9-2.

TABLE 9-2	Standard Analysis FIGURE 9-7				Thevenin Equivalent Circuit Analysis FIGURE 9-8 and FIGURE 9-9			Equivalent FIGURE 9-10
	V_{R1}	V_{R2}	V_{R3}	V_{R4}	V_{TH}	I_{SC}	R_{TH}	V_{R_L}
CALCULATED								
MEASURED								

SECTION III: Maximum Power Transfer Theorem

PART 1 (Guided Pre-lab. Calculations): Standard Analysis Techniques

Calculate the voltage drops and power dissipated by the circuit of Figure 9-11 at loaded conditions, when R1 = 10 kΩ and R_L is at 10 kΩ to achieve maximum power transfer. Then, calculate V_{RL} and P_{RL} as the R_L value is decreased to 4.7 kΩ and 2.7 kΩ and then increased to 15 kΩ and 18 kΩ.

1. With R_L at 10 kΩ, calculate the voltage drop across the 10 kΩ load resistor from:

$$V_{RL} = \frac{V_{PS} \times R_L}{R1 + R_L} \quad \text{where: } V_{PS} = 18 \text{ V}.$$

Thevenin's and Norton' Equivalent Circuits and Maximum Power Transfer — 65

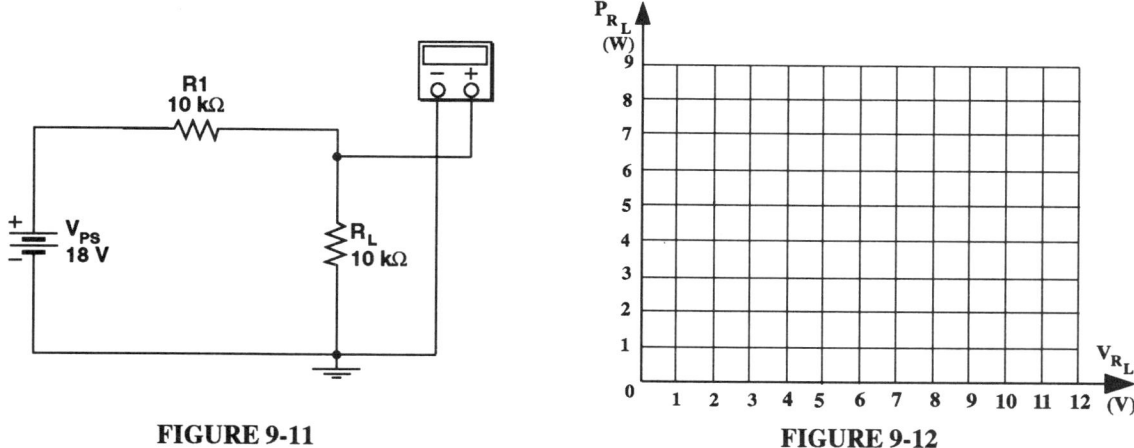

FIGURE 9-11   **FIGURE 9-12**

Then calculate the power dissipated by the 10 kΩ load resistor where: $P_{R_L} = V_{R_L}^2/R_L$.

2. In turn, connect the 2.7 kΩ, 4.7 kΩ, 15 kΩ, and 18 kΩ load resistors into the circuit. After each connection calculate the voltage dropped across each resistor where:

$$V_{R_L} = \frac{V_{PS} \times R_L}{R1 + R_L}$$

Then, calculate the power dissipated by each of the connected 2.7 kΩ, 4.7 kΩ, 15 kΩ, and 18 kΩ load resistors where: $P_{R_L} = V_{R_L}^2/R_L$.

3. Insert the calculated values, as indicated, into Table 9-3

PART B: Voltage Measurements

1. Connect, or open and select from the File menu, the circuit of Figure 9-11, where R1 = 10 kΩ and, initially, the load resistor is 10 kΩ. Measure the voltage drops across the 10 kΩ load resistor

2. In turn, connect the R1 = 10 kΩ resistor in series with each individual 2.7 kΩ, 4.7 kΩ, 15 kΩ, and 18 kΩ load resistors by changing the resistance value. Then, measure the voltage drop across each of the four load resistors.

3. Insert the measured values, as indicated, into Table 9-3

TABLE 9-3	2.7 kΩ		4.7 kΩ		10 kΩ		15 kΩ		18 kΩ	
	V_{R_L}	P_{R_L}	V_{R_L}	P_{R_L}	V_{R_L}	P_{R_L}	V_{R_L}	P_{R_L}	V_{R_L}	P_{R_L}
CALCULATED										
MEASURED		▨		▨		▨		▨		▨

PART C: Graphing the Power Transfer Curve

1. Plot the power dissipation curve, from voltage and power dissipation data of each of the load resistors of Table 9-3, in the graph of Figure 9-12. Label each 2.7 kΩ, 4.7 kΩ, 10 kΩ, 15 kΩ, and 18 kΩ point.

Questions and Problems: Basic Circuit Analysis for Electronics:133-134

CHAPTER 10
COMPLEX CIRCUIT ANALYSIS

INTRODUCTION

Circuits with two or more power sources are considered to be complex circuits that require both new and familiar techniques for analysis. Superposition, loop and nodal equations, and Thevenin's equivalent circuit are all techniques that can be used. Loop equations are based on Kirchhoff's first law stating that the sum of the voltage rises and drops in a circuit is equal to zero, while nodal equations are based on Kirchhoff's second law stating the sum of the current in and out of a node is equal to zero.

SUPERPOSITION

Superposition is a technique where one voltage source at a time is used in the analysis of a circuit such as that of Figure 10-1(a). As shown in Figure 10-1(b), V1 is the only voltage source of the circuit and V2 is shorted, which allows the circuit to be solved as a simple series-parallel circuit. The same procedure is used for the circuit of Figure 10-1(c), where V2 is the only voltage source of the circuit with V1 being shorted. The currents of each circuit are combined and the voltage drops of the two power supply circuit are then solved.

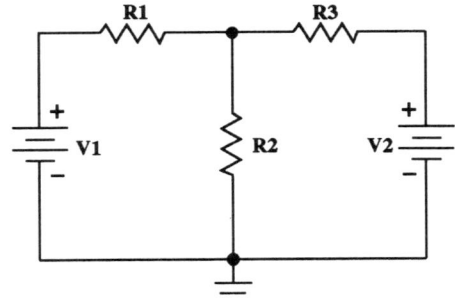
(a) Two Power Supply Circuit

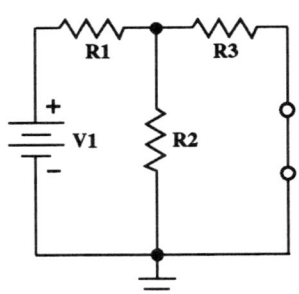
(b) V1 Single Source Circuit

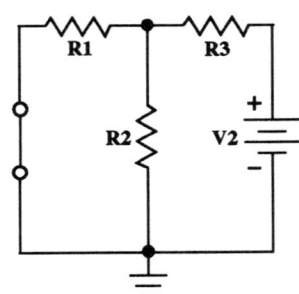
(c) V2 Single SourceCircuit

FIGURE 10-1

THEVENIN'S EQUIVALENT CIRCUIT

Thevenizing the circuit is generally accomplished by removing the R2 load resistor, as shown in Figure 10-2(a), and solving V_{R1}, V_{R2}, and V_{TH}. Then the equivialent circuit of Figure 10-2(b) is drawn and R_{TH} is solved along with the voltage across the load resistor R2. Once V_{R2} is known, the value can be substituted back into the original circuit and the remaining voltage drops and the current of the circuit are solved using Ohm's law. The calculations are shown in conjunction with the circuits of Figure 10-2, where $R2 = R_L$.

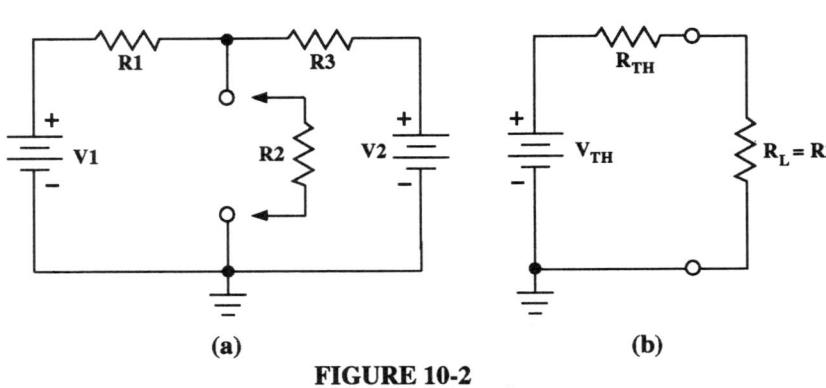

FIGURE 10-2

$$V_{R1} = \frac{(V1 - V2)R1}{R1 + R3}$$

$$V_{R3} = \frac{(V1 - V2)R3}{R1 + R3}$$

$$V_{TH} = V1 - V_{R1}, \text{ or}$$

$$V_{TH} = V2 + V_{R3}$$

$$R_{TH} = R1 \parallel R3$$

$$V_{R2} = \frac{V_{TH} \times R_L}{R_{TH} + R_L}$$

LABORATORY EXERCISE

READING ASSIGNMENT: Basic Circuit Analysis for Electronics: 135-142

EXERCISE OBJECTIVES
To become familiar with:

- Solving two power supply circuits using superposition, nodal, or loop equations.
- Solving two power supply circuits using the Thevenin equivalent analysis.

PROCEDURE

Section I: Two Power Supply Circuits
PART A. Superposition, Nodal, or Loop Equation Analysis
Pre-laboratory Calculations

1. Analyze the two power supply circuit of Figure 10-3(a). The EWB version is shown in Figure 10-3(b). In the calculations V1 is 18 V, V2 is 6 V, and circuit resistors are R1 and R2 at 2.2 kΩ and R3 at 3.3 kΩ.

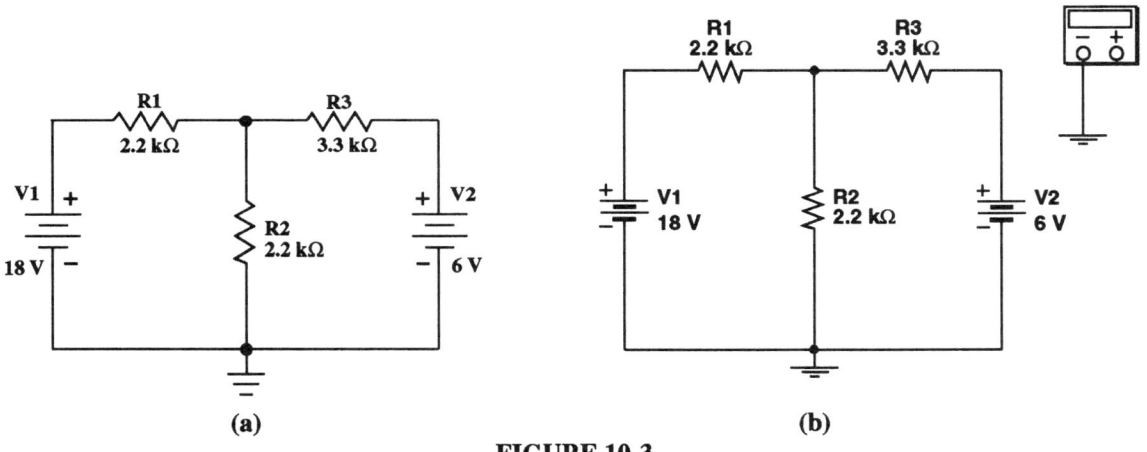

FIGURE 10-3

2. Calculate V_{R1}, V_{R2}, and V_{R3}. Use superposition, nodal, or loop equations in the analysis.

3. Calculate I_{R1}, I_{R2}, and I_{R3} using Ohm's law.

4. Insert the calculated circuit voltages and currents into Table 10-1.

Circuit Measurements

1. Connect, or open and select from the File menu, the circuit of Figure 10-3(b).

2. Measure V1, V2, and V_{R2} in the circuit of Figure 10-3(b). Measure with reference to ground.
   a. Determine the "measured" V_{R1} voltage, where $V_{R1} = V1 - V_{R2}$.
   b. Determine the "measured" V_{R3} voltage, where $V_{R3} = V_{R2} - V2$.

68 — Basic Circuit Analysis For Electronics Using Electronics Workbench®

3. Measure I_{R1}, I_{R2}, and I_{R3}, indirectly or directly.
   a. Measure indirectly, using Ohms law, where $I_{R1} = V_{R1}/R1$, $I_{R2} = V_{R2}/R2$, and $I_{R3} = V_{R3}/R3$.
   b. Measure directly, by connecting ammeters in series with the circuit resistors.

4. Insert the measured values, as indicated, into Table 10-1.

TABLE 10-1	V_{R1}	V_{R2}	V_{R3}	I_{R1}	I_{R2}	I_{R3}
CALCULATED						
MEASURED						

PART B: Theveninizing the Two Power Supply Circuit
Pre-laboratory Calculations
1. Analyze the circuit of Figure 10-4(a), where the R2 "load" resistor has been removed.

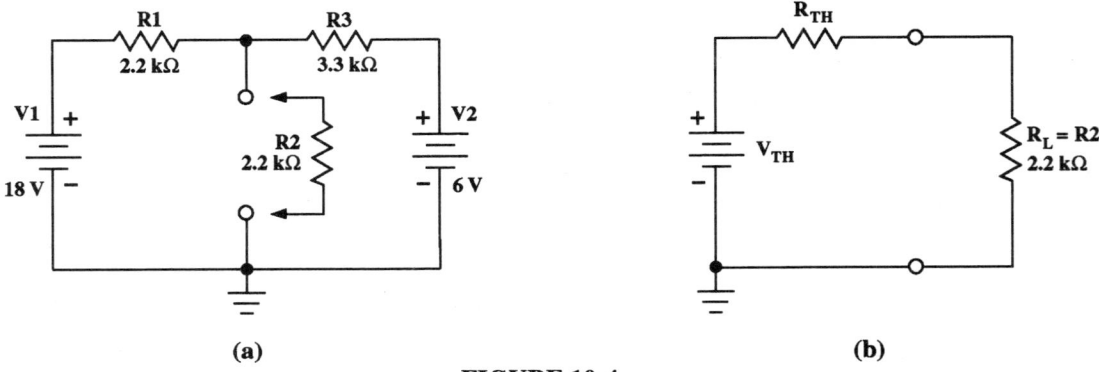

(a)　　　　　　　　　　　　　　　　　(b)

FIGURE 10-4

2. Calculate V_{TH}, R_{TH}, and V_{RL}.
   a. Calculate V_{R1} where: $V_{R1} = \dfrac{(V1 - V2)R1}{R1 + R3}$

   b. Calculate V_{R3} where: $V_{R3} = \dfrac{(V1 - V2)R3}{R1 + R3}$

   c. Calculate V_{TH} where: $V_{TH} = V1 - V_{R1}$, or $V_{TH} = V2 + V_{R3}$.

   d. Calculate R_{TH}, the theoretical parallel combination of R1 and R3 where: $R_{TH} = R1 \parallel R3$.

3. Calculate V_{RL}, the voltage across the load of the equivalent circuit of Figure 10-4(b), where:

$$V_{RL} = \dfrac{V_{TH} \times R_L}{R_{TH} + R_L}.$$

4. Insert the calculated values, as indicated, into Table 10-1.

Measuring V_{TH} and I_{SC} Directly to Solve R_{TH}
1. Construct, or open and select from the File menu, the circuit of Figure 10-5. Measure V_{TH} and I_{SC} to solve R_{TH}.
   a. Measure V_{TH}, the voltage drop across the output terminals, as shown in Figure 10-5(a).
   b. Measure $I_N = I_{SC}$ by connecting an ammeter across the output terminals, as shown in Figure 10-5(b).
   c. Use the measured V_{TH} and $I_N = I_{SC}$ values to determine R_{TH} where: $R_{TH} = V_{TH}/I_{SC}$.

Complex Circuit Analysis — 69

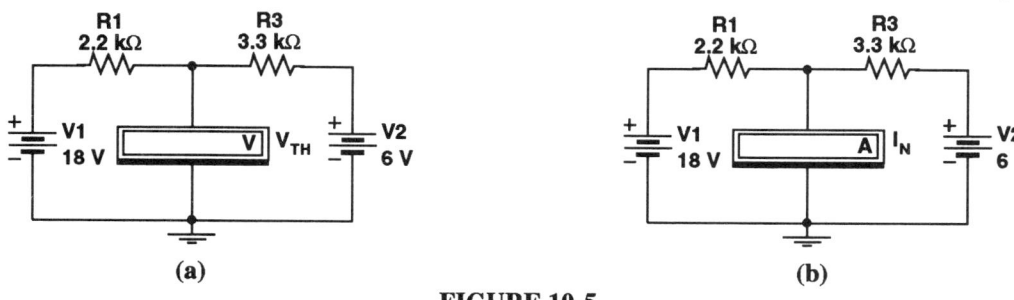

(a)   **(b)**

FIGURE 10-5

NOTE: Placing an ammeter across the output terminals effectively places the ammeter in series with V1 and R1 and V2 and R3, so the current flow through the ammeter is limited to $I_{SC} = I_{R1} + I_{R3}$.

2. Insert the measured values, as indicated, into Table 10-2.

PART C: Verifying V_{RL} Using the Equivalent Circuit
1. Construct, or open and select from the File menu, the circuits of Figure 10-6.

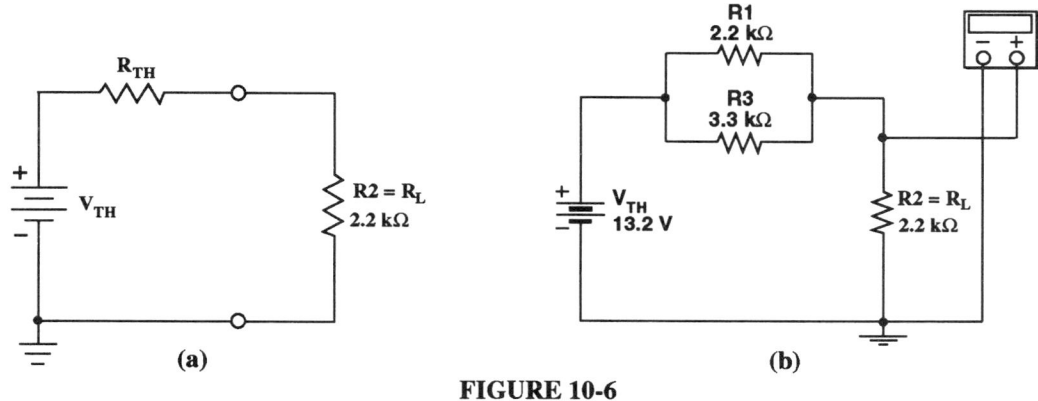

(a)   **(b)**

FIGURE 10-6

2. In the equivalent EWB circuit version shown in Figure 10-6(b), the power supply is set to the V_{TH} voltage, the R_{TH} resistance is at (R1 ∥ R3), and the load resistance is R_L = R2 at 2.2 kΩ.

3. Measure V_{RL}, the voltage across the R2 load resistor of the equivalent circuit.

4. Insert the measured values, as indicated, into Table 10-2.

TABLE 10-2	Circuit with R2 Removed							Equivalent Ckt.	
	V_{R1}	V_{R3}	V_{TH}	R_{TH}	I_{SC}	V_{RL}	V_{TH}	V_{RL}
CALCULATED					▨		▨	▨
MEASURED						▨		

SECTION II (Optional Laboratory): Two Power Supply Circuits
PART A: Analysis of Two Power Supply Circuits
Pre-laboratory Calculations
1. Analyze the two power supply circuit of Figure 10-7(a). The EWB version is shown in Figure 10-7(b). In the calculations, V1 is 15 V, V2 is 5 V, and circuit resistors are R1 at 2.2 kΩ, R2 at 3.3 kΩ, and R3 at 2.2 kΩ.

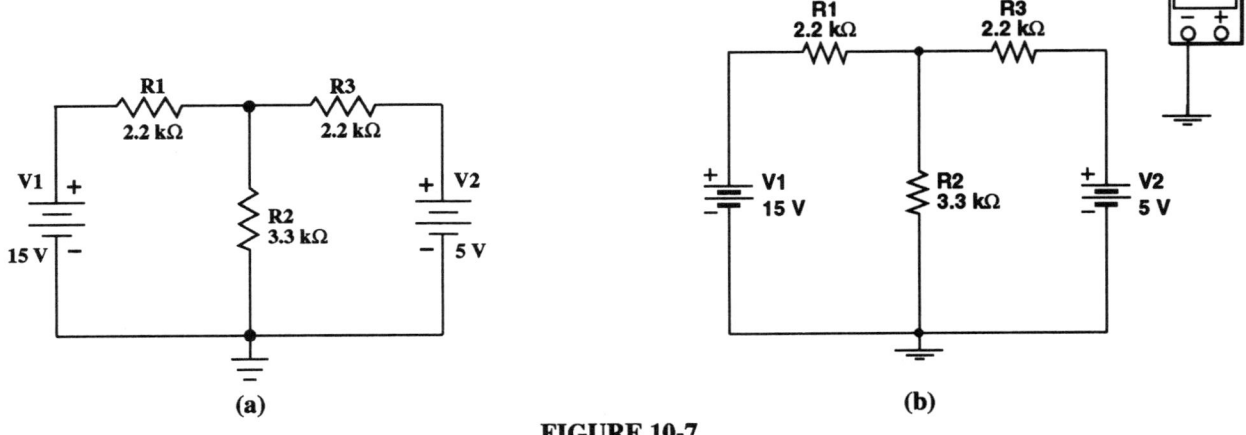

FIGURE 10-7

2. Calculate V_{R1}, V_{R2}, and V_{R3}. Use superposition, nodal, or loop equations in the analysis.

3. Calculate I_{R1}, I_{R2}, and I_{R3} using Ohm's law.

4. Insert the calculated circuit voltages and currents into Table 10-3.

Voltage Measurements
1. Connect, or open and select from the File menu, the circuit of Figure 10-7(b).

2. Measure V1, V2, and V_{R2} in the circuit of Figure 10-7(b). Measure with reference to ground.
   a. Determine the "measured" V_{R1} voltage where: $V_{R1} = V1 - V_{R2}$.
   b. Determine the "measured" V_{R3} voltage where: $V_{R3} = V_{R2} - V2$.

3. Measure I_{R1}, I_{R2}, and I_{R3} indirectly or directly.
   a. Measure indirectly, using Ohms law, where: $I_{R1} = V_{R1}/R1$, $I_{R2} = V_{R2}/R2$, and $I_{R3} = V_{R3}/R3$.
   b. Measure directly by connecting the ammeter in series with the circuit resistors.

4. Insert the measured values, as indicated, into Table 10-3.

TABLE 10-3	V_{R1}	V_{R2}	V_{R3}	I_{R1}	I_{R2}	I_{R3}
CALCULATED						
MEASURED						

PART B: Theveninizing the Two Power Supply Circuit
Pre-laboratory Calculations
1. Analyze the circuit of Figure 10-8(a) where the R2 "load" resistor has been removed.

2. Calculate V_{TH}, R_{TH}, and V_{RL}.
   a. Calculate V_{R1} where: $V_{R1} = \dfrac{(V1 - V2)R1}{R1 + R3}$.
   b. Calculate V_{R3} where: $V_{R3} = \dfrac{(V1 - V2)R3}{R1 + R3}$.

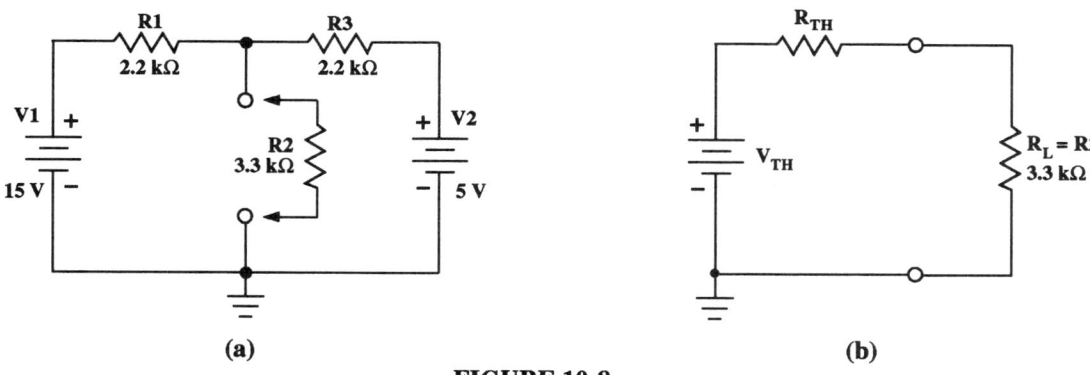

FIGURE 10-8

c. Calculate V_{TH} where: $V_{TH} = V1 - V_{R1}$ or $V_{TH} = V2 + V_{R3}$.

d. Calculate R_{TH}, the theoretical parallel combination of R1 and R3, where: $R_{TH} = R1 \parallel R3$.

3. Calculate V_{RL}, the voltage across the load of the equivalent circuit of Figure 10-8(b), where:

$$V_{RL} = \frac{V_{TH} \times R_L}{R_{TH} + R_L}.$$

4. Insert the calculated values, as indicated, into Table 10-4.

Measure V_{TH} and I_{SC} Directly to Determine R_{TH}

1. Construct, or select from the dialog box, the circuits of Figure 10-9.

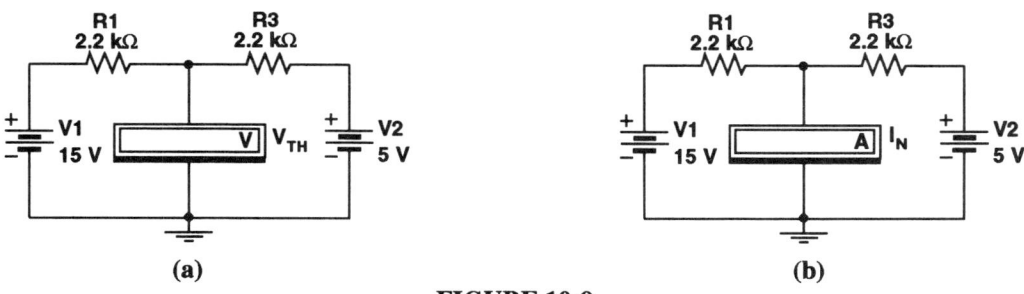

FIGURE 10-9

2. Measure V_{TH} and I_{SC} to solve R_{TH}.
   a. Measure V_{TH}, the voltage drop across the output terminals, as shown in Figure 10-9 (a).
   b. Measure $I_N = I_{SC}$ by connecting an ammeter across the output terminals, as shown in Figure 10-9 (b).
   c. Use the measured V_{TH} and $I_N = I_{SC}$ values to determine R_{TH} where: $R_{TH} = V_{TH}/I_{SC}$.

NOTE: Placing an ammeter across the output terminals effectively places the ammeter in series with the V1 and R1 and V2 and R3, so the current flow through the ammeter is limited to $I_{SC} = I_{R1} + I_{R3}$.

3. Insert the measured values, as indicated, into Table 10-4.

PART C: Verify V_{RL} Using the Equivalent Circuit

1. Construct, or select from the File menu, the circuit of Figure 10-10.

2. In the equivalent EWB circuit version shown in Figure 10-10(b), the power supply is set to the V_{TH} voltage, the R_{TH} resistance is at (R1 ∥ R3) and the load resistance is $R_L = R2$ at 3.3 kΩ.

3. Measure V_{RL}, the voltage across the load resistor of the equivalent circuit.

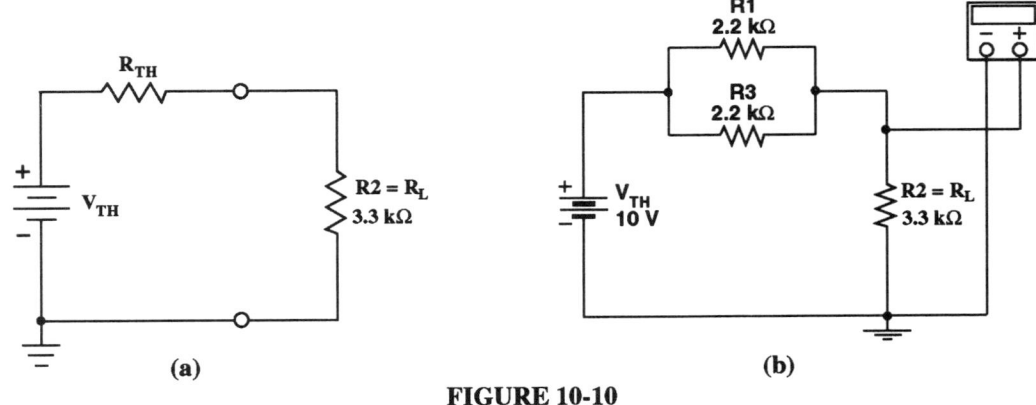

FIGURE 10-10

4. Insert the measured values, as indicated, into Table 10-4.

TABLE 10-4	Circuit with R2 Removed						Equivalent Ckt.	
	V_{R1}	V_{R3}	V_{TH}	R_{TH}	I_{SC}	V_{R_L}	V_{TH}	V_{R_L}
CALCULATED					/////		/////	/////
MEASURED						/////		

Questions and Problems: Basic Circuit Analysis for Electronics: 146-147

CHAPTER 11
CIRCUIT LOADING EFFECTS

INTRODUCTION
Connecting a load resistor to a circuit, even a high resistive load, causes current to flow through the load and voltages in the circuit to be redistributed. Three examples of loading effect are examined below.

METER LOADING
Voltmeters have a relatively high input resistance, typically 10 MΩ in the EWB program. So, while voltmeters provide reasonably close calculated to measured values across low value resistors, loading effect across high value resistors occurs even with high input resistance voltmeters. So for the circuit of Figure 11-1, if the circuit has low resistor values the equations for low resistance resistors are quite acceptable but for high value resistors the resistive value of the meter should be taken into account.

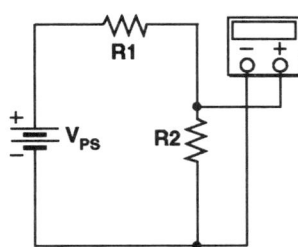

$$V_{R2} = \frac{V_{PS} \times R2}{R1 + R2}$$

$$V_{R1} = V_{PS} - V_{R2}$$

Low Resistance Resistor Equations

$$V_{R2} = \frac{V_{PS} \times (R2 \parallel RM)}{R1 + (R2 \parallel RM)}$$

$$V_{R1} = V_{PS} - V_{(R2 \parallel RM)}$$

High Resistance Resistor Equations

FIGURE 11-1: Circuit under test.

BLEEDER CIRCUIT LOADING
The bleeder circuit is used to provide a dc voltage that does not vary greatly between unloaded or loaded circuit condition. To achieve this condition, the current in the bleeder circuit (bleeder current) is typically much larger than the current into the load. In the analysis, the first formula is used to solve VA, then VA = V_{R5} is used to solve V_{R2}, and VB = V_{R4}. The equations are shown in conjunction with Figure 11-2.

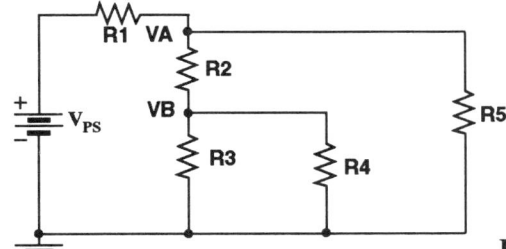

$$VA = \frac{V_{PS} \times [(R2 + [R3 \parallel R4]) \parallel R5]}{R1 + [(R2 + [R3 \parallel R4]) \parallel R5]}$$

$$V_{R2} = \frac{VA \times R2}{R2 + (R3 \parallel R4)}$$

$$VB = V_{R3} = V_{R4} = \frac{VA \times (R3 \parallel R4)}{R2 + (R3 \parallel R4)}$$

FIGURE 11-2

BRIDGE CIRCUIT LOADING
The loaded bridge circuit of Figure 11-3 can be solved using loop equations, nodal equations, or delta wye conversion. However, if only the voltage drop across the load is required the Thevenin analysis is used.

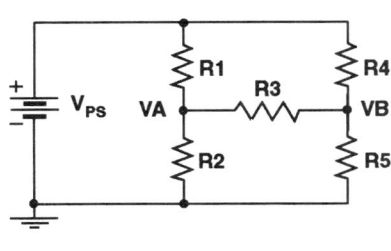 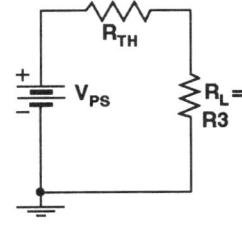

$$VA = \frac{V_{PS} \times R2}{R1 + R2} \quad \text{and} \quad VB = \frac{V_{PS} \times R5}{R4 + R5}$$

$$V_{TH} = VA - VB$$

$$R_{TH} = (R1 \parallel R2) + (R4 \parallel R5)$$

$$V_{RL} = \frac{V_{TH} \times R3}{R_{TH} + R3} = \frac{V_{TH} \times R_L}{R_{TH} + R_L}$$

FIGURE 11-3

LABORATORY EXERCISE

READING ASSIGNMENT: Basic Circuit Analysis for Electronics: 148-156

EXERCISE OBJECTIVES
To become familiar with:

- Meter loading of high resistance circuits.
- Loaded bleeder circuit.
- Loaded bridge circuits.

PROCEDURE

SECTION I: Two resistor Series Circuit
PART A: Laboratory Calculations
1. Analyze the circuit of Figure 11-4(a). The EWB version is shown in Figure 11-4(b).

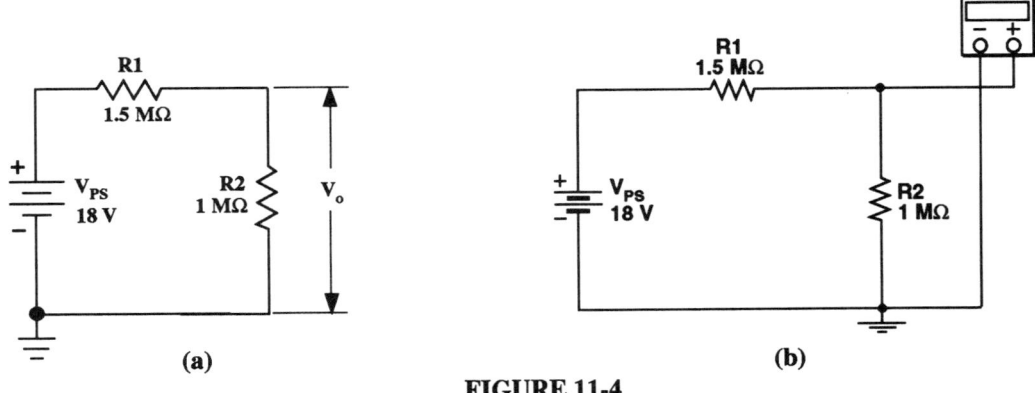

FIGURE 11-4

2. Solve V_{R1} and V_{R2} using Ohm's law or voltage divider equations, where R1 is 1.5 MΩ and R2 is 1 MΩ.

3. Then, solve V_{R1} and V_{R2}, using in the calculations the loading effect of the internal resistance of the voltmeter RM at 10 MΩ, where:

$$V_{R1} = \frac{V_{PS} \times R2}{R1 + RP} \quad \text{and} \quad V_{R2} = V_{RP} = \frac{V_{PS} \times R2}{R1 + RP} \quad \text{where: } RP = R2 \parallel RM$$

4. Insert the calculated values into Table 11-1.

PART B: Voltage Measurements
1. Construct, or open and select from the File menu, the circuit of Figure 11-4(b)
   a. Measure $V_{R2} = V_{RP}$ directly, measured with respect to reference ground.
   b. Measure V_{R1} indirectly, knowing the V_{R2} measured voltage, where: $V_{R1} = V_{PS} - V_{R2}$.
   c. Use the measured V_{R1} voltage to solve the current through R1 where: $I_{R1} = V_{R1}/R_1$.
   d. Use the $V_{R2} = V_{RP}$ and I_{R1} values to solve RP from: $RP = V_{RP}/I_{R1}$.

e. Use the R2 and RP values to solve RM where: $RM = \dfrac{R2 \times RP}{R2 - RP}$.

2. Insert the measured values, as indicated, into Table 11-1.

TABLE 11-1	Coded Resistor		Meter Loaded Conditions				
	V_{R1}	V_{R2}	V_{R1}	$V_{RP} = V_{RP}$	I_{R1}	RP	RM
CALCULATED							
MEASURED							

SECTION II: Bleeder Resistor Circuit
PART A: Unloaded Circuit Calculations
1. Calculate voltage drops of the unloaded bleeder resistor circuit of Figure 11-5(a).

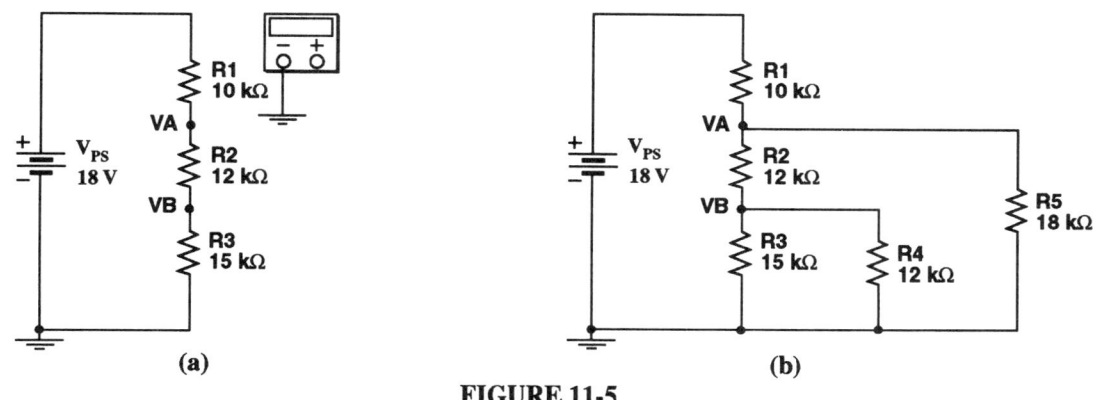

FIGURE 11-5

2. Solve the circuit voltage drops using Ohm's law or voltage divider equations.

3. Insert the calculated values, as indicated, into Table 11-2.

PART B: Voltage Measurements (Unloaded Circuit)
1. Construct, or open and select from the File menu, the circuit of Figure 11-5(a).

   a. Measure V_{PS}, VA, and VB, measured with respect to reference ground.

   b. Use the measured V_{PS}, VA, and VB = V_{R3} voltages to determine V_{R1} and V_{R2} indirectly.

2. Insert the measured values, as indicated, into Table 11-2.

PART C: loaded Circuit Calculations
1. Calculate voltage drops of the loaded bleeder resistor circuit of Figure 11-5(b). Use V_{PS} at 18 V and the resistances of the circuit in the calculations.

2. Solve the circuit voltage drops using Ohm's law or voltage divider equations.

3. Insert the calculated values into Table 11-2.

PART D: Voltage Measurements (Loaded Circuit)
1. Construct, or open and select from the File menu, the circuit of Figure 11-5(b)

   a. Measure V_{PS}, VA, and VB, measured with respect to reference ground.

   b. Use the measured V_{PS}, VA = V_{R5}, and VB = V_{R3} = V_{R4} voltages to determine V_{R1} and V_{R2} indirectly.

2. Insert the measured values, as indicated, into Table 11-2

TABLE 11-2	Unloaded Circuit				R4 and R5 Resistor Loaded Circuit			
	V_{R1}	V_{R2}	$V_{R3} = V_B$	V_A	V_{R1}	V_{R2}	$V_{R3} = V_{R4} = V_B$	$V_{R5} = V_A$
CALCULATED								
MEASURED								

SECTION III: Bridge Circuit

PART A: Circuit Calculations

1. Calculate the voltage drops of the unloaded circuit of Figure 11-6(a), where the 18 kΩ load resistor R3 has been disconnected.

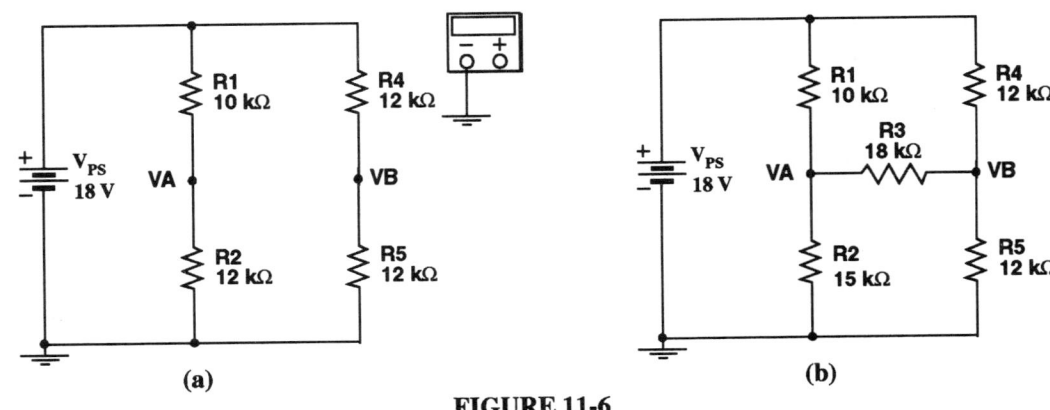

FIGURE 11-6

2. Solve the unloaded voltage values of V_{R1}, $VA = V_{R2}$, V_{R4}, and $VB = V_{R5}$. Use basic circuit analysis techniques.

3. Calculate V_{TH} and R_{TH}.
   a. Solve V_{TH} where: $V_{TH} = VA - VB$.
   b. Solve R_{TH} where: $R_{TH} = (R1 \parallel R2) + (R4 \parallel R5)$.

4. For the loaded circuit of Figure 11-6(b), calculate $V_{RL} = V_{R3}$ where: $V_{RL} = \dfrac{V_{TH} \times R_L}{R_{TH} + R_L}$.

5. Insert the calculated values into Table 11-3.

PART B: Voltage Measurements

1. Construct, or open and select from the File menu, the circuit of Figure 11-6(a).
   a. Measure V_{PS}, VA, and VB. Measure with respect to reference ground.
   b. Use the measured V_{PS}, VA = V_{R2}, and VB = V_{R5} voltages to solve V_{R1} and V_{R4} indirectly.

2. Construct, or open and select from the File menu, the loaded circuit of Figure 11-6(b):
   a. Measure V_{PS}, VA, and VB. Measure with respect to reference ground.
   b. Use the measured V_{PS}, VA = V_{R2}, and VB = V_{R5} voltages to solve V_{R1}, V_{R4}, and $V_{R3} = V_{RL}$ indirectly.

PART C: Equivalent Circuit Construction and Measurement

1. Construct, or select from the File menu, the equivalent circuit of Figure 11-7(b), which is the expaned version of the equivalent circuit of Figure 11-7(a).

2. Adjust the power supply, if the circuit is constructed, to the calculated V_{TH} voltage.

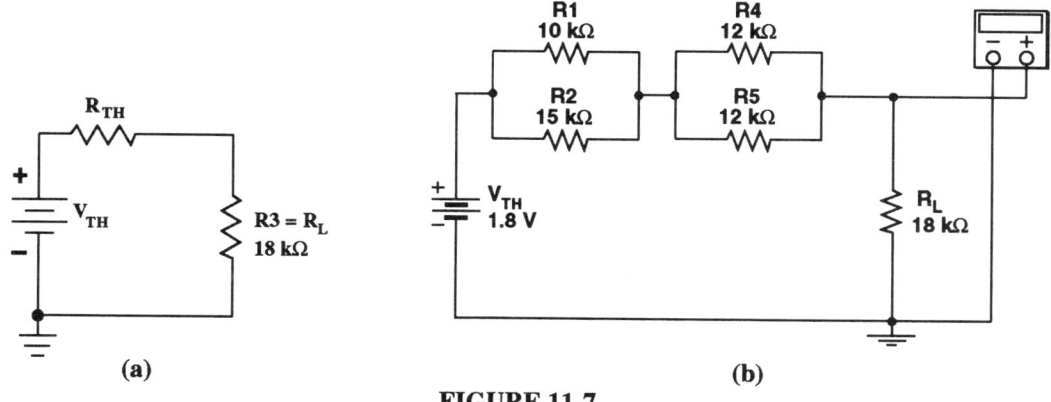

FIGURE 11-7

3. Measure V_{R_L}, the voltage across the R3 = R_L = 18 kΩ resistor.

4. Insert the measured values into Table 11-3.

TABLE 11-3	FIGURE 11-6					FIGURE 11-7		
	V_{R2} = VA	R_{TH}	V_{R5} = VB	V_{TH}	$V_{R3}=V_{RL}$	V_{TH}	R_{TH}	V_{RL}
CALC.						▨		
MEAS.		▨		▨			▨	

Questions and Problems: Basic Circuit Analysis for Electronics: 160

CHAPTER 12
CAPACITORS IN DC CIRCUITS

INTRODUCTION

Capacitors, unlike resistors, perform their functions only when there is a change in voltage. This limits their use in dc circuits to those times when the circuit is switched on or off or the voltage levels are changed. During these times capacitors can be charged, store a charge, or discharge current back through the circuit. They act as a kind of temporary battery to store voltage and oppose changes in voltage.

The voltage across the capacitor of the circuit of Figure 12-1(a) charges exponentially, as shown in the universal charge curve of Figure 12-1(b). So, the voltage rises with time and reaches a maximum voltage charge condition (ideally) in five time constants. This is where the capacitor reaches 63.2% of the forcing voltage in the first time constant, and then 63.2% of the remaining forcing voltages for each of the remaining significant four time constants. Also, the time constant for the RC circuit is the product of the resistance in ohms and the capacitance in farads measured in seconds. So, $\tau = RC$ is the time constant and V_{max} is effectively achieved after 5 time constants. The voltage across the charging capacitor can also be solved using the exponential equation $V_C = V_{PS}(1 - e^{-t/\tau}) = V_{PS}(1 - e^{-t/RC})$.

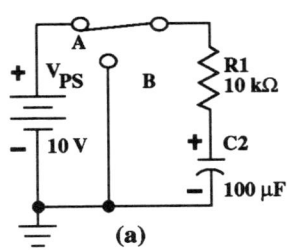

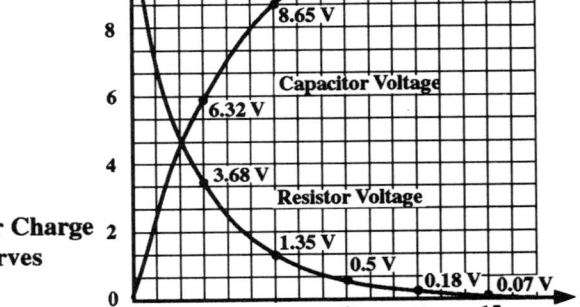

(b) Capacitor-Resistor Charge and Discharge Curves

FIGURE 12-1

RC CIRCUIT

The advantage of using the standard exponential formula in the calculations of the charging voltage across the capacitor in a circuit is that the voltage can be solved for any time. Also, if the capacitors in the circuit are in parallel, the total capacitance value equals $C_T = C1 + C2$, as shown in Figure 12-2(a). If the capacitors are in series the total capacitance value equals $C_T = C1 \parallel C2$, as shown in Figure 12-2(b). Then, in the circuit of Figure 12-2(c), the circuit is converted to a Thevenin equivalent circuit which allows the voltage developed across the output capacitor to be analyzed at any time.

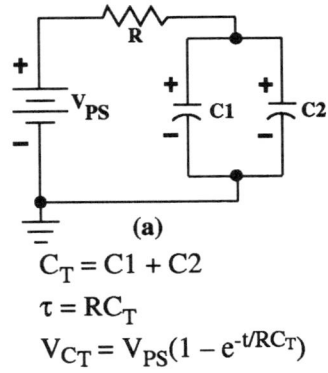

(a)
$C_T = C1 + C2$
$\tau = RC_T$
$V_{CT} = V_{PS}(1 - e^{-t/RC_T})$

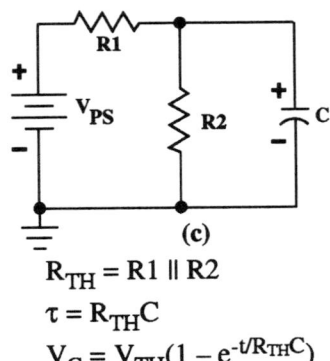

(b)
$C_T = C1 \parallel C2$
$\tau = RC_T$
$V_{CT} = V_{PS}(1 - e^{-t/RC_T})$

(c)
$R_{TH} = R1 \parallel R2$
$\tau = R_{TH}C$
$V_C = V_{TH}(1 - e^{-t/R_{TH}C})$

FIGURE 12-2

Capacitors in DC Circuits — 79

LABORATORY EXERCISE

READING ASSIGNMENT: Basic Circuit Analysis for Electronics: 161-171

EXERCISE OBJECTIVES
To become familiar with:

- Basic RC time constant circuits.

- RC network analysis with capacitors in parallel.

- RC network analysis with capacitors in series.

- RC network analysis using the Thevenin techniques.

PROCEDURE

SECTION I: Basic Resistor-Capacitor Circuit
PART A: Series Resistor-Capacitor Circuit
Pre-laboratory Calculations
1. Analyze the basic RC circuit of Figure 12-3(a). In the calculations use V_{PS} at 15 V, a 100 kΩ resistor, and a 100 µF capacitor. The EWB version is shown in Figure 12-3(b). Calculate the time constant (τ), and the voltage developed across the capacitor C at each 10 second time interval, 10 through 50 seconds, after the switch makes contact and the circuit is activated.

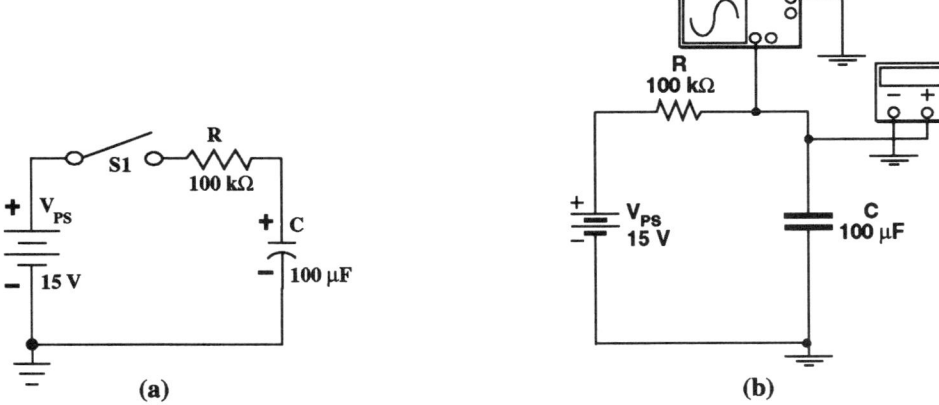

FIGURE 12-3

2. Calculate the RC time constant of the circuit of Figure 12-3(a), where $\tau = RC$, R = 100 kΩ, and C is measured.

3. Calculate V_{C1}, the voltage developed across the C1 capacitor at time intervals of 10 s, 20 s, 30 s, 40 s, and 50 s.
   a. Solve for V_C at a time (t) of 10 s, where $V_C = V_{PS}(1 - e^{-t/\tau}) = V_{PS}(1 - e^{-t/RC})$.
   b. Then repeat the calculations of V_C at the remaining time intervals of 20 s, 30 s, 40 s, and 50 s.

4. Insert the calculated values, as indicated, into Table 12-1.

Measurements

1. Construct, or open and select from the File menu, the circuit of Figure 12-3(b), where the switch is initially closed (circuit activated), so that the voltage across the capacitor begins at 0.0 V.

2. Use the voltmeter to measure the output voltage (V_o) across the output capacitor, where the oscilloscope is connected to achieve the needed exponential response on the dc voltmeter.
   a. Measure the initial 0.0 V developed across capacitor C at 0.0 seconds, prior to the switch being activated.
   b. Then activate the circuit and measure V_o, measured across C, at each of the 10 s, 20 s, 30 s, 40 s, and 50 s intervals. (Use the Pause/Resume button at each of the 10 s intervals.)

3. Insert the measured values, as indicated, into Table 12-1.

TABLE 12-1	Voltage Developed Across C After					
	0.0 s	10 s	20 s	30 s	40 s	50 s
CALCULATED						
MEASURED						

NOTE: Measured values on the voltmeter, for this circuit connection, will be slightly lower than the calculated values.

Plotting the Graph.

1. In the graph of Figure 12-4, plot the calculated V_o versus each of the 10 s, 20 s, 30 s, 40 s and 50 s time intervals.

2. Then on the same graph of Figure 12-4, superimpose the measured plot, referencing V_o at each of the 10 s through 50 s time intervals.

3. Label each exponential curve, ideally using different colors, so the curves can be identified.

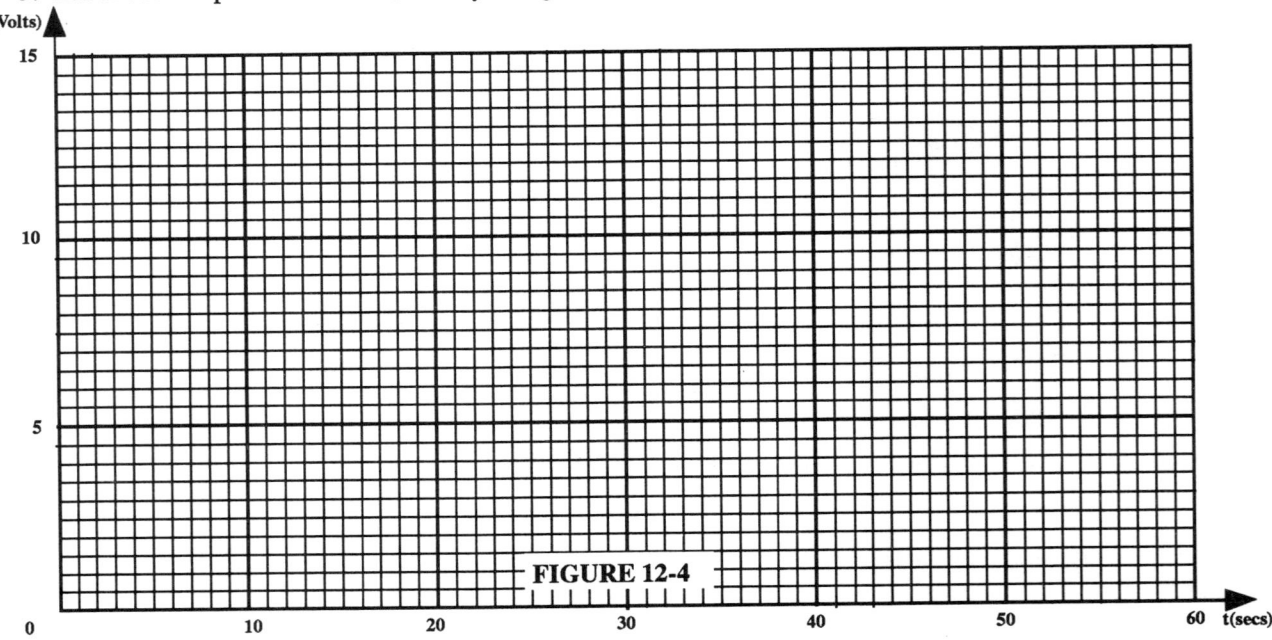

FIGURE 12-4

PART B: Series Capacitor-Resistor Circuit
Pre-laboratory Calculations

1. Analyze the basic CR circuit of Figure 12-5. In the calculations the power supply is set to 15 V, the resistor

is 100 kΩ, and the capacitor is 100 μF. The EWB version is shown in Figure 12-5(b). Calculate the time constant (τ), and the voltage developed across R at time intervals of 10 through 50 seconds, after the circuit is activated.

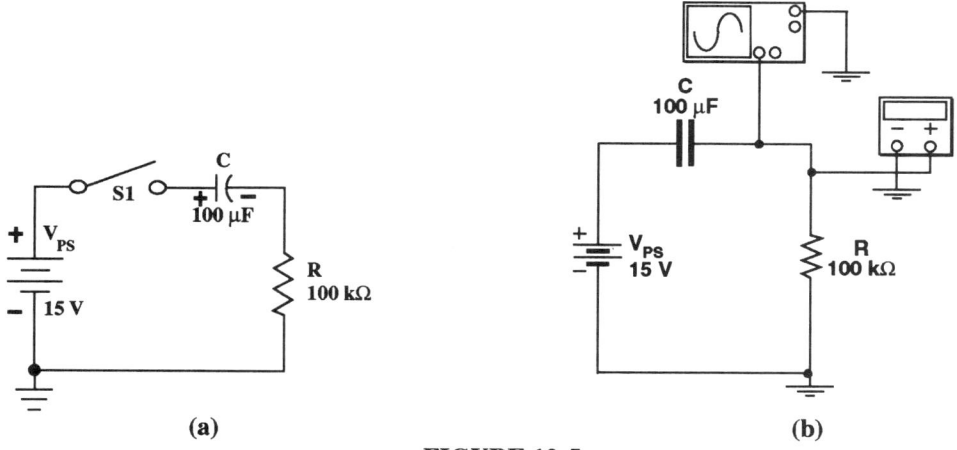

FIGURE 12-5

2. Calculate the CR time constant of the Figure 12-5 circuit where: τ = RC, R = 100 kΩ, and C is 100 μF.

3. Calculate V_R, the voltage developed across resistor R at time intervals of 10 s(seconds), 20 s, 30 s, 40 s, and 50 s.
   a. Solve for V_R at a time (t) of 10 s, where $V_R = V_{PS}(e^{-t/RC})$.
   b. Then repeat the calculations of V_C at the remaining time intervals of 20 s, 30 s, 40 s, and 50 s.

4. Insert the calculated values, as indicated, into Table 12-2.

Measurements

1. Construct, or open and select from the File menu, the circuit of Figure 12-5(b), where the voltmeter is set to dc and the output voltage is monitored across the resistor R.

2. Use the voltmeter, where the oscilloscope is connected to achieve the exponential response on the dc voltmeter.
   a. Measure the initial 0.0 V developed across capacitor C, at 0.0 seconds, prior to the switch being opened.
   b. Then activate the circuit and measure V_o across R at each of the 10 s, 20 s, 30 s, 40 s, and 50 s intervals.

NOTE: The voltage across resistor R should be approximately equal to V_{PS} just when the circuit is activated.

3. Insert the measured values, as indicated, into Table 12-2.

TABLE 12-2	Voltage Developed Across R After					
	0.0 s	10 s	20 s	30 s	40 s	50 s
CALCULATED						
MEASURED						

Plotting the Graph.

1. In the graph of Figure 12-6, plot the calculated V_o versus each of the 10 s, 20 s, 30 s, 40 s, and 50 s time intervals.

2. Then, on the same graph of Figure 12-6, superimpose the measured plot, referencing V_o at each of the 10 s through 50 s time intervals.

3. Label each of the exponential curves, ideally using different colors, so the curves are easily identified.

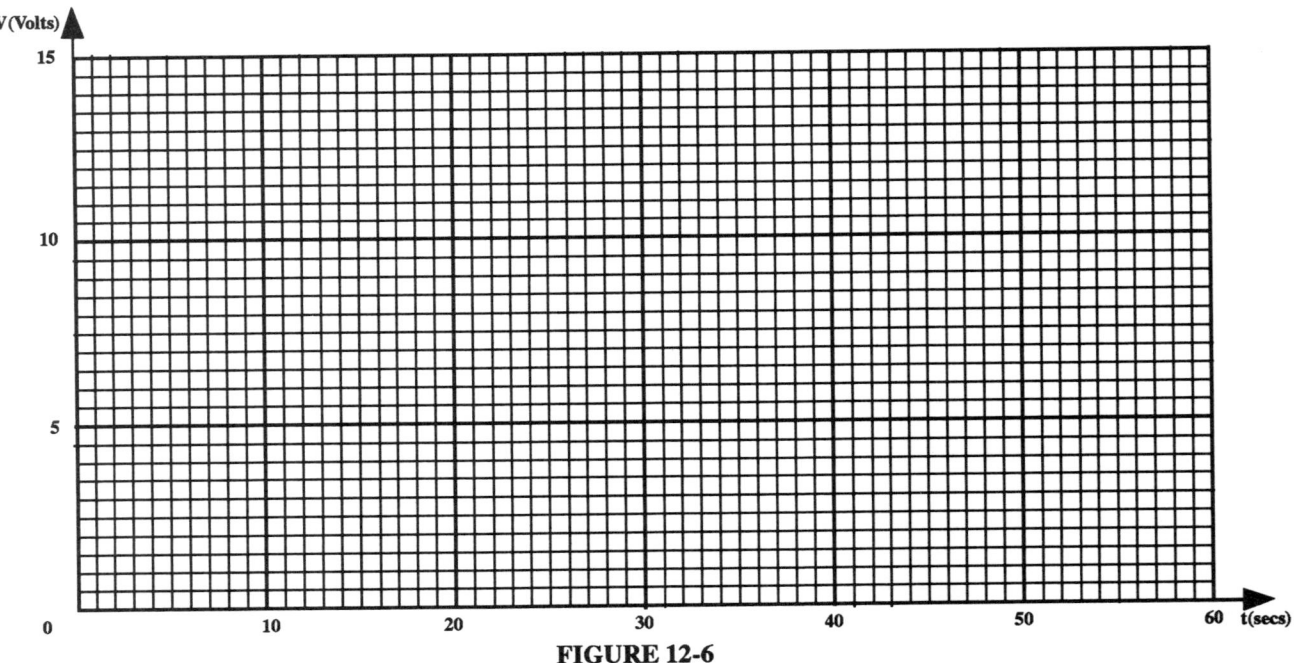

FIGURE 12-6

SECTION II: Series Resistor with both Parallel and Series Capacitors
PART A: Serial Resistance and Parallel Capacitors Circuit
Pre-laboratory Calculations

1. Analyze the R(C1 ∥ C2) circuit of Figure 12-7. In the calculations the power supply is 15 V, the resistor 100 kΩ, and the parallel capacitors are each 100 µF. Calculate the time constant (τ), and the voltage developed across the parallel C1 and C2 capacitors at time intervals of 20 through 100 seconds, after the switch makes contact and the circuit is activated.

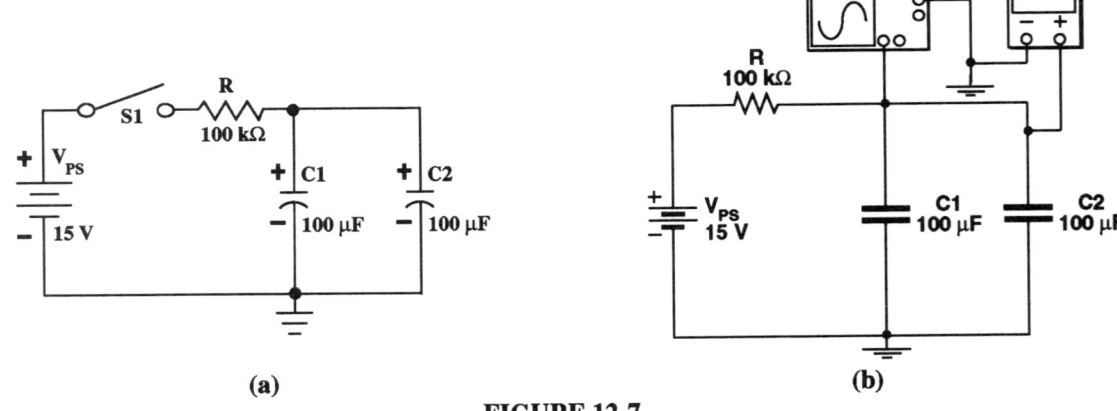

FIGURE 12-7

2. Calculate the RC_T time constant of the circuit of Figure 12-7 where: $\tau = RC_T$, R1 = 100 kΩ, and C1 and C2 are each 100 µF.

NOTE: Capacitors in parallel are solved like resistors in series where: $C_T = C1 + C2$.

3. Calculate V_{C_T}, the voltage developed across the parallel C1 and C2 capacitors, at time intervals of

Capacitors in DC Circuits — 83

20 s(seconds), 40 s, 60 s, 80 s, and 100 s.

a. Solve for V_{CT} at a time (t) of 20 s, where: $V_C = V_{PS}(1 - e^{-t/RC_T})$.

b. Then repeat the calculations of V_{CT} at the remaining time intervals of 40 s, 60 s, 80 s, and 100 s.

4. Insert the calculated values, as indicated, into Table 12-3.

Measurements

1. Construct, or open and select from the File menu, the circuit of Figure 12-7(b), where the voltmeter is set to dc and the output voltage is monitored across the parallel capacitors.

2. Use the voltmeter, where the scope is connected to achieve the exponential response on the dc voltmeter.
   a. Measure the 0.0 V developed across the parallel C1 and C2 capacitors at 0.0 seconds, prior to the circuit being activated.
   b. Then activate the circuit and measure V_o across the parallel C1 and C2 capacitors at each of the 20 s, 40 s, 60 s, 80 s, and 100 s intervals.

3. Insert the measured values, as indicated, into Table 12-3.

TABLE 12-3	Voltage Developed Across C1 in Parallel with C2 After					
	0.0 s	20 s	40 s	60 s	80 s	100 s
CALCULATED						
MEASURED						

Plotting the Graph.

1. In the graph of Figure 12-8, plot the theorertical V_o versus each of the 20 s, 40 s, 60 s, 80 s, and 100 s time intervals.

2. Then, on the same graph of Figure 12-8, superimpose the measured plot, referencing V_o at each of the 20 s through 100 s time intervals.

3. Label each of the exponential curves, ideally using different colors, so the curves are easily identified.

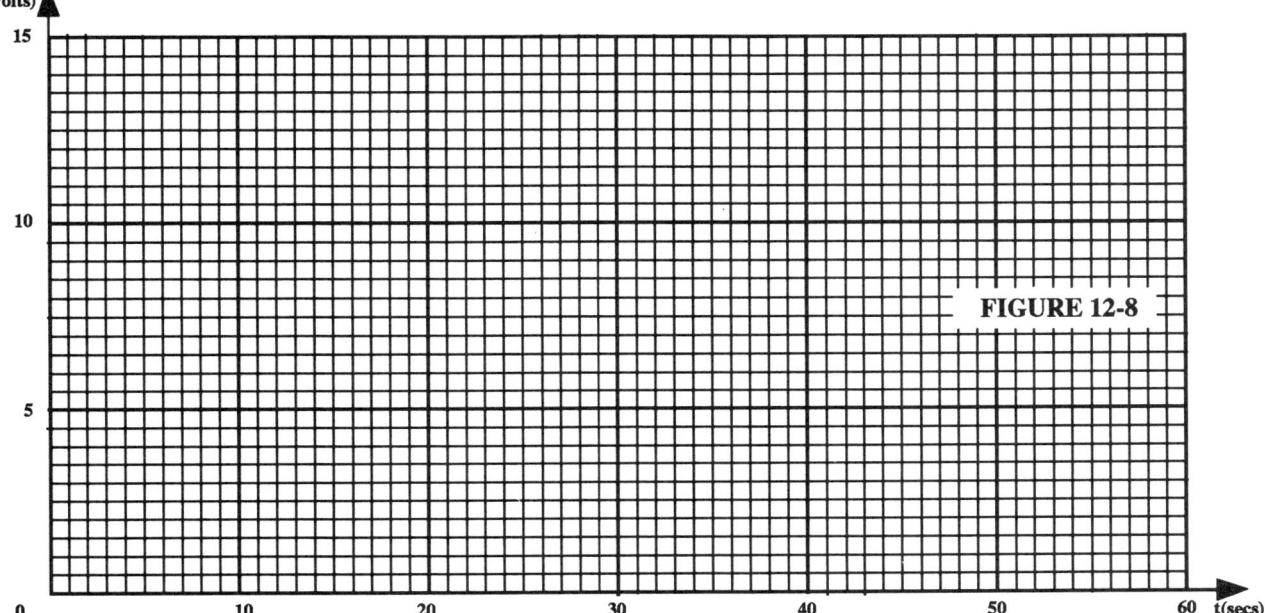

FIGURE 12-8

PART B. Resistance and Series Capacitors Circuit
Pre-Laboratory Calculations

1. Analyze the RC_T circuit of Figure 12-9. In the calculations the power supply is set to 15 V, the resistor is 100 kΩ, and the series capacitors are each 100 µF. Calculate the time constant (τ) and the voltage developed across the series C1 and C2 capacitors at time intervals of 5 through 25 seconds, after the switch makes contact and the circuit is activated.

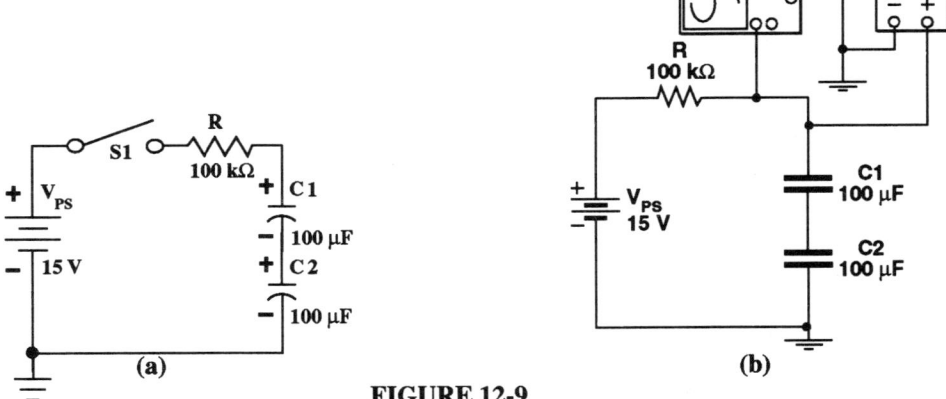

FIGURE 12-9

2. Calculate the RC_T time constant of the circuit of Figure 12-9, where $\tau = RC_T$, R = 100 kΩ, and series C1 and C2 are individually measured.

NOTE: Capacitors in series are solved like resistors in parallel so $C_T = \dfrac{1}{1/C1 + 1/C2}$.

3. Calculate V_{CT}, the voltage developed across series capacitors C1 and C2, at time intervals of 5 s (seconds), 10 s, 15 s, 20 s, and 25 s.
   a. Solve for V_{CT} at a time (t) of 5 s, where $V_C = V_{PS}(1 - e^{-t/RC_T})$.
   b. Then repeat the calculations of V_{CT} at the remaining time intervals of 10 s, 15 s, 20 s, and 25 s.

4. Insert the calculated values, as indicated, into Table 12-4.

Measurements

1. Construct, or open and select from the File menu, the circuit of Figure 12-9(b), where the voltmeter is set to dc and the output voltage is monitored across the series capacitors.

2. Use the voltmeter, where the oscilloscope is connected to achieve the exponential response on the dc meter.
   a. Measure the 0.0 V developed across the series C1 and C2 capacitors at 0.0 seconds, prior to the circuit being activated.
   b. Then activate the circuit and measure V_o across the series C1 and C2 capacitors at each of the 5 s, 10 s, 15 s, 20 s, and 25 s intervals.

3. Insert the measured values, as indicated, into Table 12-4.

TABLE 12-4	Voltage Developed Across C1 in Series with C2 After					
	0.0 s	5 s	10 s	15 s	20 s	25 s
CALCULATED						
MEASURED						

Plotting the Graph.

1. On the graph of Figure 12-10, plot the calculated V_o versus each of the 5 s, 10 s, 15 s, 20 s, and 25 s time intervals.

2. Then, on the same graph of Figure 12-10, superimpose the measured plot, referencing V_o at each of the 5 s through 25 s time intervals.

3. Label each of the exponential curves, ideally using different colors, so the curves are easily identified.

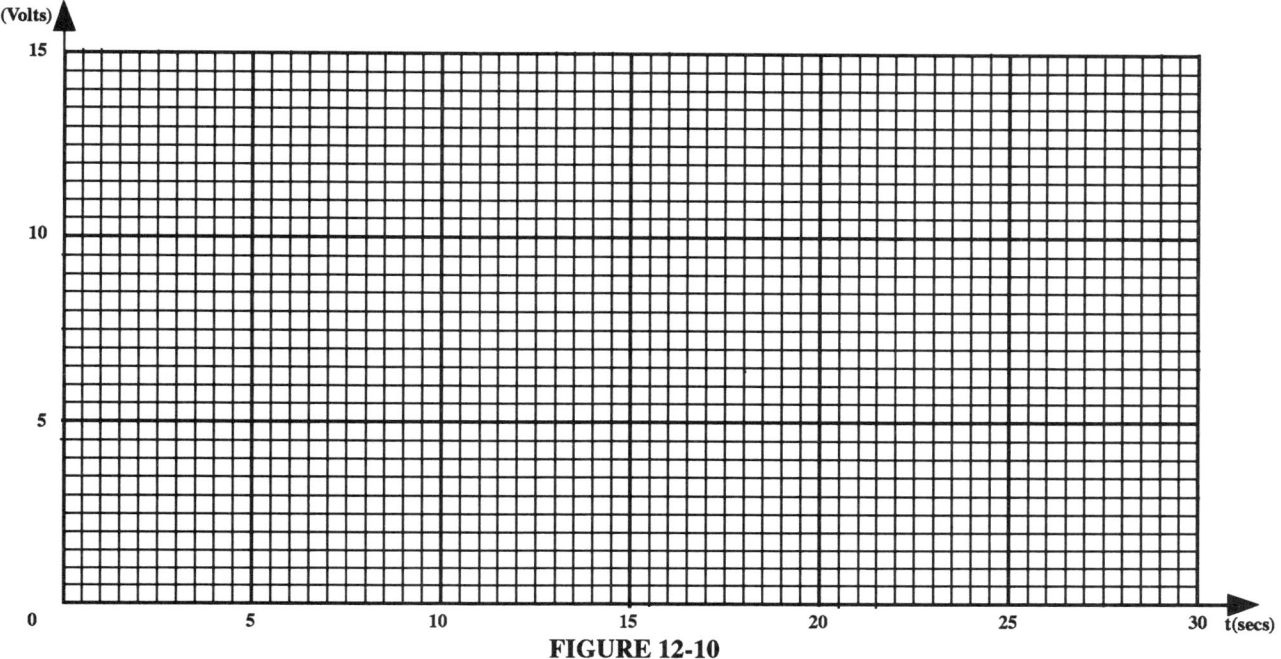

FIGURE 12-10

SECTION III: Series-Parallel Resistor and Capacitor Circuit
PART A: Pre-laboratory Calculations

1. Analyze the circuit of Figure 12-11(a). In the calculations use the power supply at 15 V, R1 at 100 kΩ, R2 at 150 kΩ, and the capacitor at 100 μF. Find V_{TH} and R_{TH} in solving the circuit. Calculate the time constant (τ), and the voltage developed across the parallel R2 resistor and C capacitor at time intervals of 6 through 30 seconds, after the switch makes contact and the circuit is activated.

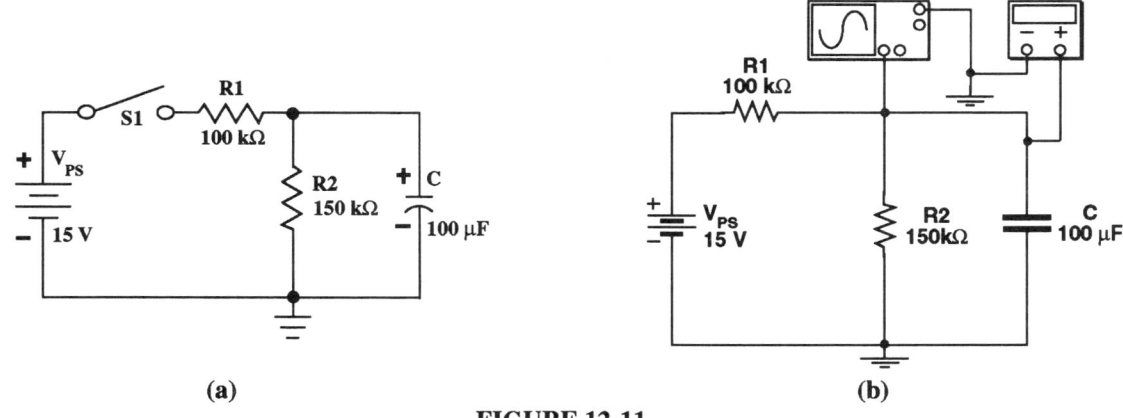

FIGURE 12-11

2. Calculate the $R_{TH}C$ time constant of the circuit of Figure 12-11 where: $\tau = R_{TH}C$, R1 = 100 kΩ, R2 = 150 kΩ, and C is 100 μF.

NOTE: $V_{TH} = \dfrac{V \times R2}{R1 + R2}$ and $R_{TH} = R1 \parallel R2$.

86 — Basic Circuit Analysis For Electronics Using Electronics Workbench®

3. Calculate the voltage developed across the parallel R2 and C components at time intervals of 6 s, (seconds), 12 s, 18 s, 24 s, and 30 s.
   a. Solve for V_o at a time (t) of 6 s, where $V_o = V_{TH}(1 - e^{-t/R_{TH}C})$.
   b. Then repeat the calculations of V_o at the remaining time intervals of 12 s, 18 s, 24 s, and 30 s.

4. Insert the calculated values, as indicated, into Table 12-5.

PART B: Measurements

1. Construct, or open and select from the File menu, the circuit of Figure 12-11(b), where the voltmeter is set to dc and the output voltage is monitored across the series capacitors.

2. Use the voltmeter, where the scope is connected to achieve the exponential response on the dc voltmeter.
   a. Measure the 0.0 V developed across the parallel R2 and C components, at 0.0 seconds, prior to the circuit being activated.
   b. Then activate the circuit and measure V_o across (R2 ∥ C) at each of the 6 s through 30 s intervals.

3. Insert the measured values, as indicated, into Table 12-5.

TABLE 12-5	Voltage Developed Across C in Parallel with R2 After					
	0.0 s	6 s	12 s	18 s	24 s	30 s
CALCULATED						
MEASURED						

PART C: Plotting the Graph.

1. In the graph of Figure 12-12, plot the calculated V_o versus each of the 6 s through 30 s time intervals.

2. Then, on the same graph of Figure 12-12, superimpose the measured plot, referencing V_o at each of the 6 s through 30 s time intervals.

3. Label each of the exponential curves, ideally using different colors, so the curves are easily identified.

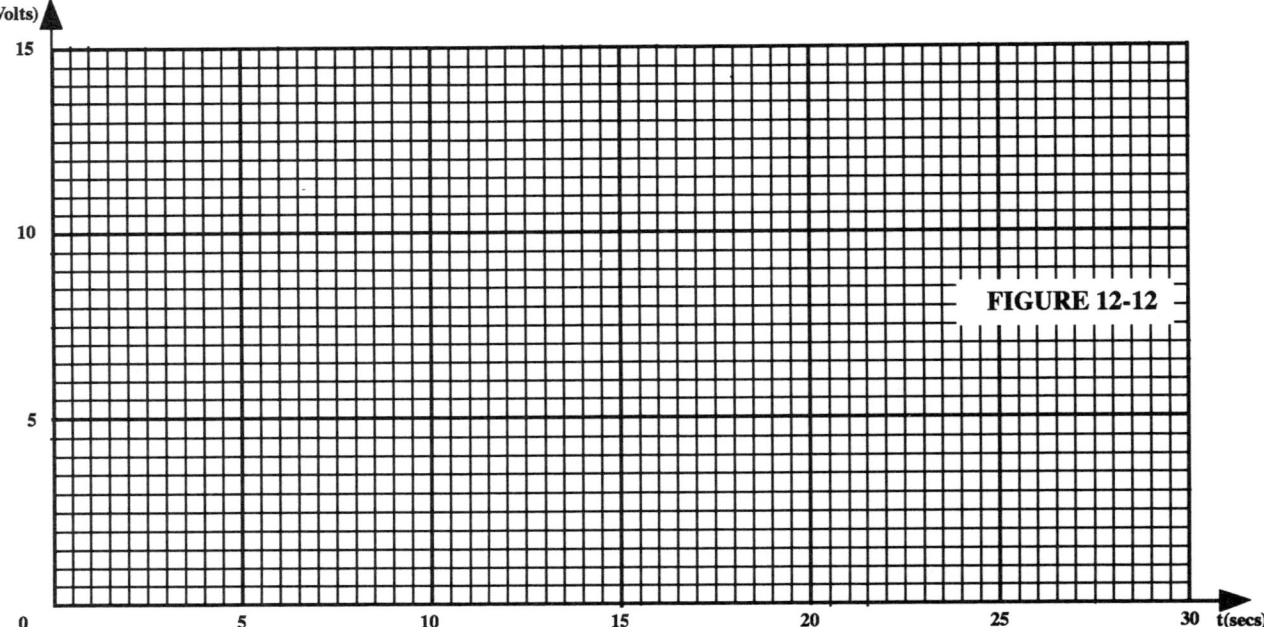

FIGURE 12-12

Questions and Problems: Basic Circuit Analysis for Electronics: 180

CHAPTER 13
INTRODUCTION TO AC SINE WAVE ANALYSIS AND OSCILLOSCOPE MEASUREMENTS

INTRODUCTION

In the EWB program the sine wave voltage is provided by the function generator or the ac voltage source. The function generator provides a peak-to-peak voltage that is typically measured on the oscilloscope, as shown in Figure 13-1(a). The ac voltage source provides a rms voltage that is typically measured on the voltmeter, as shown in Figure 13-1(b). However, if the output voltage of the ac rms voltage source is measured on the scope it will be peak-to-peak voltage and will be $2\sqrt{2}$ times larger than the rms input voltage value. Likewise if the peak-to-peak voltage is measured on a voltmeter it will be $2\sqrt{2}$ times smaller than the peak-to-peak voltage input voltage.

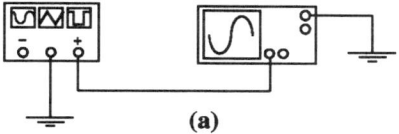
(a)
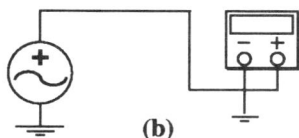
(b)

FIGURE 13-1

LABORATORY GENERATOR

To obtain the ac source drag it from the Sources bin and double click on the icon. When the dialog box appears, make the appropriate changes and click on accept. An example as shown in Figure 13-2(a) is where the rms sine wave voltage is set to 3 V rms, the frequency to 1 kHz, and the duty cycle at 50%.

To obtain a sine wave peak-to-peak voltage drag the function generator from the Instruments bin. It is enlarged, as shown in Figure 13-2(b), by double clicking on the function generator icon, which allows a sine, square, or triangle wave shape to be selected. Then, the frequency, duty cycle, amplitude, and offset voltage can be set. So, to choose the sine wave, click on the sine wave, set the frequency to 1 kHz, and set the amplitude at 6 Vp-p. However in order to obtain the 6 Vp-p on the oscilloscope, set the amplitude of the generator to 3 V, which indicates a peak voltage setting, or 6 Vp-p. The sine wave is chosen by clicking on it and the frequency is set by typing in 1 and using the arrow to select kHz. The default duty cycle at 50% and the offset voltage at 0.0 V are the settings generally used and are not changed.

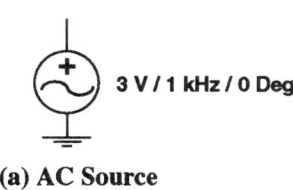
(a) AC Source

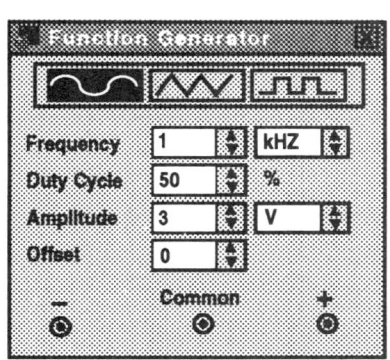
(b) Function Generator

FIGURE 13-2

OSCILLOSCOPES

Oscilloscopes are the most versatile of all electronic measuring instruments. Scopes can be used to measure either dc voltage or ac waveshapes where, for ac, both the amplitude and the frequency of the ac waveshape are visually measured. So, any distortion to the waveshapes is easily detected. The main controls of the dual-trace oscilloscope, shown in Figure 13-3 and used in the measurement of the vertical peak-to-peak voltage in volts/div, are the vertical inputs and controls for channel A and for channel B. The two channels allow signal voltages to be compared, for instance, the input and output of a circuit. The horizontal sec/div control measures the time in seconds horizontally on the screen. The time base, which is the same for both signal inputs of the dual-trace oscilloscope, has a horizontal position control and a switchable time base. Once the time in seconds of one cycle is known, the frequency of the waveshape displayed can be solved from $f = 1/t$. Other controls on scopes are the trigger control and a switch that allows the input signal to be referenced to ground.

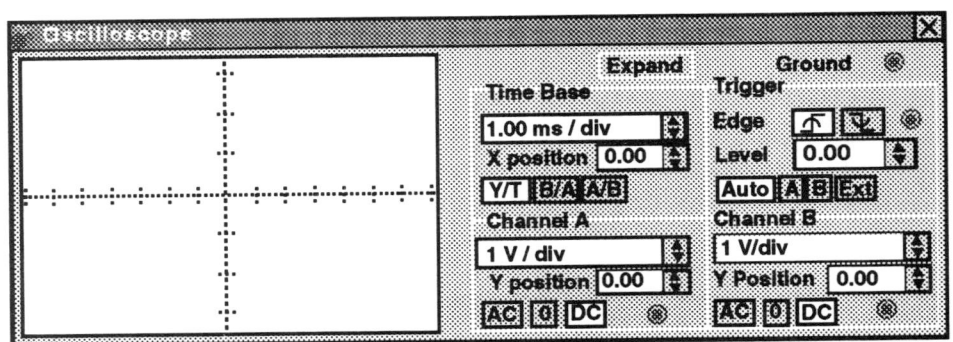

FIGURE 13-3

Bode Plotter

The bode plotter is used to graph the frequency response of circuits such as low pass, high pass, band reject and bandpass filters. In future course work, the Bode plotter will be used to provide the frequency response of amplifier circuits. The connection for the Bode plotter is shown in Figure 13-4(a), where the input is connected to the generator, the output is connected to the output of the circuit, and the grounds are mutually connected. The enlarged version is provided in Figure 13-4(b), where the mode is set to magnitude and the typical vertical axis setting are to set to log, I (initial) at – 40 dB, and F (final) at 0.0 dB. While a typical horizontal axis setting is to set to log I (initial) at 1 Hz and F (final) at 10 MHz.

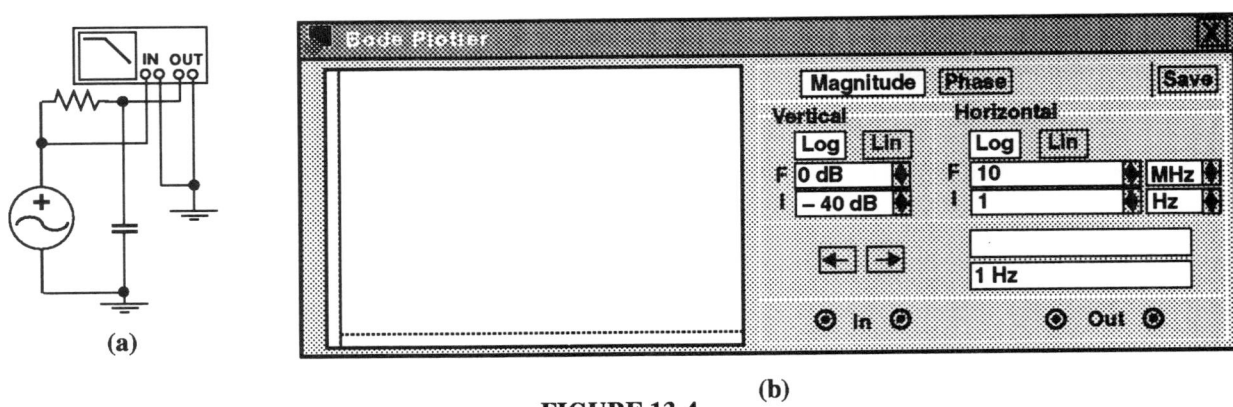

FIGURE 13-4

Once the circuit is activated and the frequency response plotted, the vertical cursor, which is on the left hand side of the Bode plotter display, can be dragged to a new location. Typically, it is dragged to the – 3 dB point location, where both the dB and frequency can be read on the control panel. (The cursor can also be moved in incremental steps using the arrows on the control panel.)

Introduction to AC Sine Wave Analysis and Oscilloscope Measurements —89

LABORATORY EXERCISE

READING ASSIGNMENT: Basic Circuit Analysis for Electronics: 181-190.

EXERCISE OBJECTIVES
To become familiar with:

- Basic ac oscilloscope measurements.
- Basic dc oscilloscope measurements.
- Applying Vp-p and V rms sine wave voltages

Procedure

SECTION I: Function Generator and the Oscilloscope
PART A: Generator and Oscilloscope Set-ups
1. The initial generator settings are shown in Figure 13-5(b).
   a. The frequency of the generator is set to 1 kHz, to sine wave, and to 50% duty cycle.
   b. The amplitude is set to 3 V peak which provides 6 Vp-p.

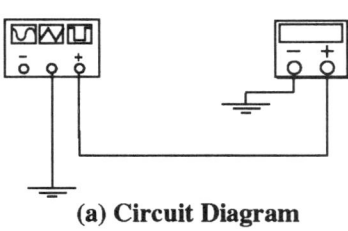

(a) Circuit Diagram

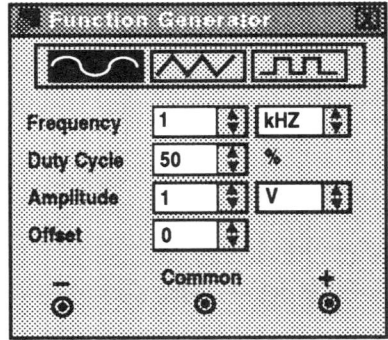

(b) Function Generator

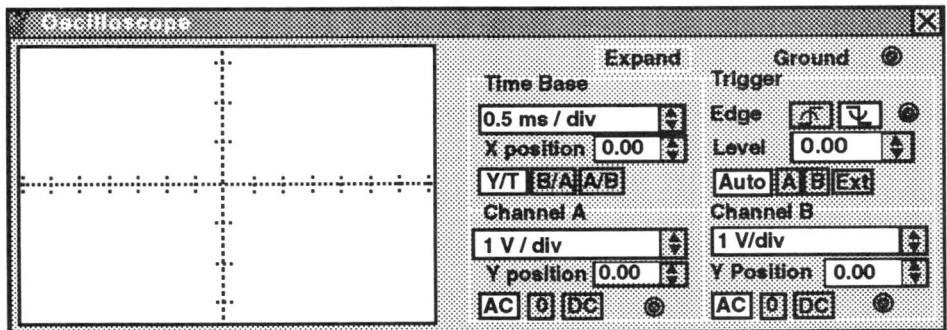

(c) Oscilloscope

FIGURE 13-5

2. The initial oscilloscope settings are shown in Figure 13-5(c). The amplitude (vertical) for Channel A is set to 1 V/div and the time base (horizontal) is set to 0.5 ms/div.
   a. The Y/T is selected because it allows the vertical (Y position) setting that is measured in volts to be compared to the horizontal (X position) setting that is measured in time.
   b. The trigger level is set to 0.0 and automatic (AUTO) and to positive (EDGE) triggering.
   c. Channel A is switched to AC mode and the vertical and horizontal controls to 0.0 to center the trace.
   d. Channel B is set to 0.0.

PART B: Measuring the Amplitude of the Sine Wave

Open and select from the File menu the circuit diagram of Figure 13-5(a), or connect the circuit diagram.. In the connection, the signal generator is connected to the Channel A vertical input of the scope.

The amplitude of the 6 Vp-p sine wave will be measured on three different (volts/div) control settings. Set and maintain the sine wave of the generator frequency at 1 kHz and the time base of the scope at 0.5 ms/div for all three volts/div settings. Click on the scope icon to enlarge the scope. Then, activate the circuit by clicking on the power switch. For better accuracy, click on Expand and adjust the horizontal sliding cursors to determine the exact positive peak voltage, the negative peak voltage, and the peak-to-peak voltage.

1. Scope Setting: 1 V/div and 0.5 ms/div
   a. Set the scope to AC mode and make sure the vertical (volts/div) control of channel A is set to 1 V/div and the horizontal (time/div) control is set to 0.5 ms/div.
   b. Make sure the function generator is set to 1 kHz, sine wave mode, 3 V peak voltage, and the peak-to-peak sine wave fills six vertical divisions on the scope.
   c. Determine the peak-to-peak amplitude of the sine wave by recording the number of vertical divisions the sine wave occupies and knowing the (volts/div) control setting where: Vp-p = div × volts/div.

2. Scope setting: 2 V/div and 0.5 ms/div
   a. Switch the vertical (volts/div) on the oscilloscope to 2 V/div while maintaining the horizontal (time/div) control at 0.5 ms/div.
   b. Determine the peak-to-peak amplitude of the sine wave by recording the number of vertical divisions the sine wave occupies and knowing the (volts/div) control setting where: Vp-p = div × volts/div.

3. Scope setting: 5 V/div and 0.5 ms/div
   a. Switch the vertical (volts/div) on the oscillosscope to 5 V/div while maintaining the horizontal (time/div) control at 0.5 ms/div.
   b. Determine the peak-to-peak amplitude of the sine wave by recording the number of vertical divisions the sine wave occupies and knowing the (volts/div) control setting where: Vp-p = div × volts/div.

4. Record the number of divisions and Vp-p, for each of the voltage settings, in Table 13-1.

PART C: Measuring Time of the Sine wave

The time of one cycle of a sine wave will be measured on three different (time/div) control settings. Set and maintain the sine wave of the generator frequency at 1 kHz and 6 Vp-p. Keep the amplitude of the scope at 1 V/div for all three time/div settings. Click on the scope icon to enlarge the scope. Then, activate the circuit by clicking on the power switch. For better accuracy, click on Expand and adjust the horizontal sliding cursors to determine the time of one complete cycle.

1. Scope Setting: 0.2 ms/div and 1 V/div
   a. Set the horizontal (time/div) control on the scope to 0.2 ms/div, the vertical (volts/div) control to 1 V/div, and the amplitude at 6 Vp-p.
   b. Set the frequency on the generator to 1 kHz and monitor the scope, where one cycle of the sine wave fills five (5) horizontal divisions on the screen.

Introduction to AC Sine Wave Analysis and Oscilloscope Measurements — 91

    c. After activating the circuit, determine the time of one complete sine wave cycle by recording the number of horizontal divisions the sine wave occupies and knowing the t/div control knob setting. Therefore: t(sec) = divs × time/div. For better accuracy, adjust cursor #1 to determine T1 and cursor #2 to determine T2, where the time of one cycle is measured directly on the expanded scope from: t(sec) = T2 − T1.

2. Scope Setting: 0.5 ms/div and 1 V/div
   a. Switch the horizontal (time/div) control on the scope to 0.5 ms/div. Maintain the vertical (volts/div) control at 1 V/div and the amplitude at 6 Vp-p.
   b. Determine the time of one cycle of the sine wave by recording the number of horizontal divisions the sine wave occupies and the (time/div) control setting where: t(sec) = divs × time/div.

3. Scope Setting: 1 ms/div and 1 V/div
   a. Switch the horizontal (time/div) control on the scope to 1 ms/div, Maintain the vertical (volts/div) control at 1 V/div and the amplitude at 6 Vp-p.
   b. Determine the time of one cycle of the sine wave by recording the number of horizontal divisions the sine wave occupies and the (time/div) control setting where: t(sec) = divs × time/div.

4. Use the time to determine the frequency where: f = 1/t.

5. Record the number of divisions and the time for each of the time/div settings in Table 13-1.

TABLE 13-1	Amplitude (volts/div)			Time (secs/div)			Frequency
	1 V/div	2 V/div	5 V/div	0.2 ms/div	0.5 ms/div	1 ms/div	f(Hz)
CALCULATED	div	div	div	div	div	div	
MEASURED	Vp-p	Vp-p	Vp-p	ms	ms	ms	

PART D: Combined Voltage, Time, and Frequency Measurements

An oscilloscope is not only capable of measuring the amplitude and time of each cycle of a sine wave, but it can be used to determine the frequency, once the time is known, from: f = 1/t. So, for each sine wave, along with the oscilloscope setting provided, determine the Vp-p, time, and frequency.

1. Scope Setting: 0.2 V/div and 0.5 ms/div
   a. On the scope set the vertical (volts/div) control to 0.2 V/div and the horizontal (time/div) control to 0.5 ms/div. Then, set the amplitude of the generator to 1.2 Vp-p (0.6 Vpeak) and the frequency at 1 kHz.
   b. Determine the number of vertical divisions and the amplitude where: Vp-p = divs × volts/div.
   c. Determine the number of horizontal divisions and the time where: time = divs × time/div.
   d. Determine the frequency where: f(Hz) = 1/t(sec).

2. Scope Setting: 0.5 V/div and 0.2 ms/div
   a. On the scope set the vertical (volts/div) control to 0.5 V/div and the horizontal (time/div) control to 0.2 ms/div. Then, set the amplitude of the generator to 1 Vp-p (0.5 Vpeak) and the frequency at 1.25 kHz.
   b. Determine the number of vertical divisions and the amplitude where: Vp-p = divs × volts/div.
   c. Determine the number of horizontal divisions and the time where: time = divs × time/div.
   d. Determine the frequency where: f(Hz) = 1/t(sec).

3. Scope Setting: 1 V/div and 50 µs/div
   a. On the scope set the vertical (volts/div) control to 1 V/div and the horizontal (time/div) control to 50 µs/div. Then, set the amplitude of the generator to 4 Vp-p (2 Vpeak) and the frequency at 10 kHz.
   b. Determine the number of vertical divisions and the amplitude where: Vp-p = divs × volts/div.
   c. Determine the number of horizontal divisions and the time where: time = divs × time/div.
   d. Determine the frequency where: f(Hz) = 1/t(sec).

4. Record the measured vertical and horizontal divisions, the (Vp-p), time (t), and frequency (f), as indicated, into Table 13-2.

TABLE 13-2	1. Amplitude (0.2 V/div) Time (0.5 ms/div)			2. Amplitude (0.5 V/div) Time (0.2 ms/div)			3. Amplitude (2 V/div) Time (50 µs/div)		
Measured	Vert. Div.			Vert. Div.			Vert. Div.		
Measured	Horz. Div.			Horz. Div.			Horz. Div.		
Calculated	Vp-p			Vp-p			Vp-p		
Calculated	t ms	f	kHz	t ms	f	kHz	t µs	f	kHz

SECTION II: DC Voltage Measurements on the oscilloscope
PART A: Positive DC Voltage Measurements

An oscilloscope is not only capable of measuring ac voltage, but it can also measure dc voltages. If a dc voltage is being measured, a reference voltage is established and the scope is set to the dc mode. The positive voltage connection is shown in Figure 13-6(a). Either open and select Figure 13-6(a) from the File menu, or connect the circuit.

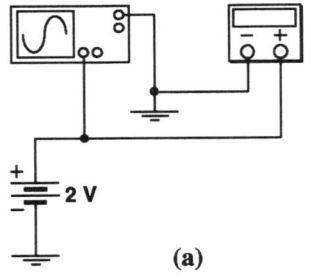
(a)

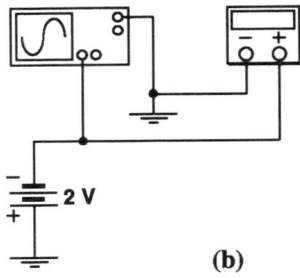
(b)

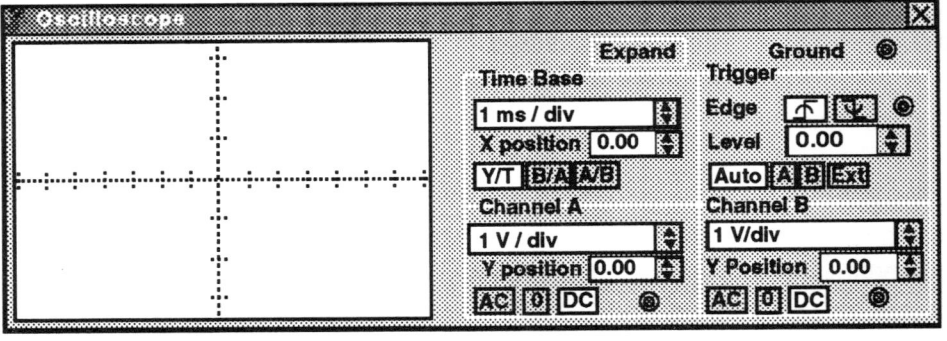
(c) Oscilloscope Settings

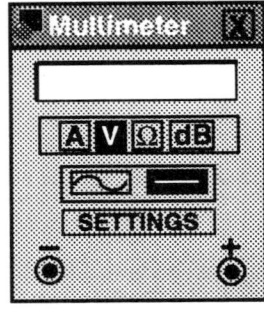
(d) Multimeter Settings

FIGURE 13-6

Introduction to AC Sine Wave Analysis and Oscilloscope Measurements — 93

1. Initial Scope Setting.
   a. On the scope, set the Channel A vertical (volts/div) control to 2 V/div and the horizontal (time/div) control to 1 ms/div.
   b. Make sure the scope is set to DC mode and the Y position at 0.0 V to center the reference in measuring the plus-and-minus dc voltage on the scope.

2. Connect the dc power supply to both the dc volt meter and the oscilloscope, as shown in Figure 13-6(a). The dc voltage will be set at 2 V initially. Then, in turn, the voltage will be set in at 4 V and 6 V.

3. Vary the dc power supply to 2 V.
   a. Monitor the dc voltage on the dc voltmeter. Click on dc (—) to make the measurement.
   b. On the scope, set to 0 or AC mode to establish the 0.0 V reference voltage. Then, switch to the DC mode to provide the positive dc voltage measurement.
   c. Record the number of vertical divisions the trace deflected and in what direction from the 0.0 V reference center line. Then determine the dc voltage on the scope where: V dc = divs × volts/div.

4. Adjust the dc power supply to 4 V and then repeat the procedure for 6 V.
   a. Monitor the dc voltage on the dc meter.
   b. On the scope, set to 0 or AC mode to establish the reference voltage. Then switch to the DC mode to provide the positive dc voltage measurement.
   c. Record the number of vertical divisions the trace deflected and in what direction from the 0.0 V reference center line. Then determine the dc voltage on the scope where: V dc = divs × volts/div.

5. Insert the positive voltage value, as indicated, into Table 13-3.

PART B: Negative DC Voltage Measurements

Either open and select Figure 13-6(b) from the File menu or construct the circuit. Reverse the dc power supply leads as shown. Initially, set to 0 or AC mode to establish the 0.0 V reference center line. Connect the dc power supply to both the oscilloscope and the dc voltmeter, as shown in Figure 13-6(b). The dc voltage will be set at – 2 V initially. Then, it will be set, in turn, at – 4 V and – 6 V. The scope settings are maintained at 2 V/div and 1 ms/div.

1. Vary the dc power supply to – 2 V.
   a. Monitor the dc voltage on the dc meter. Click on the dc (—) to make the measurement.
   b. On the scope, set to 0 or AC mode to establish the reference voltage, and then to the DC mode to provide the negative dc voltage measurement.
   c. Record the number of vertical divisions the trace deflected and in what direction from the 0.0 V reference center line. Then determine the dc voltage on the scope, where V dc = divs × volts/div.

2. Reset the dc power supply to – 4 V and then repeat the procedure for – 6 V.
   a. Monitor the dc voltage on the dc meter. Click on the dc (—) to make the measurement.
   b. On the scope, set to 0 or AC mode to establish the reference voltage and to the DC mode to provide the negative dc voltage measurement.
   c. Record the number of vertical divisions the trace deflected and in what direction from the 0.0 V reference center line. Then determine the dc voltage on the scope, where V dc = divs × volts/div.

3. Insert both the negative voltage value, as indicated, into Table 13-3.

TABLE 13-3	Positive Supply			Negative Supply		
	2 V	4 V	6 V	– 2 V	– 4 V	– 6 V
CALCULATED	div	div	div	div	div	div
MEASURED	V	V	V	V	V	V

SECTION III: Basic AC Circuit Analysis

PART A: V rms Voltage

1. Calculate the circuit voltage drops of the circuit of Figure 13-7(a), where the input voltage (V_{in}) of the ac generator is set to 3 V rms and 1 kHz.

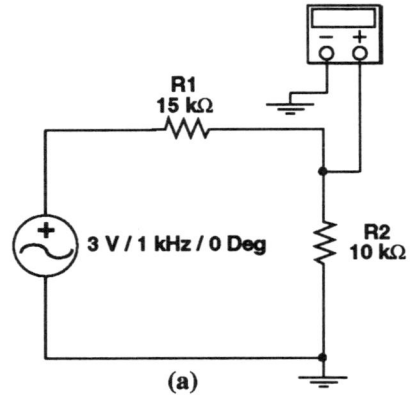

(a) (b)

FIGURE 13-7

2. Calculate V_{R1} and V_{R2}.
   a. Calculate V_{R1} where $V_{R1} = V_{in} \times R1/(R1+ R2)$.
   b. Calculate V_{R2} where $V_{R2} = V_{in} \times R2/(R1+ R2)$.

3. Open and select Figure 13-7(a) from the File menu, or connect the circuit. Measure V_{R2}, as shown in Figure 13-7(a). Then, determine the voltage drop across R1 indirectly where: $V_{R1} = V_{in} - V_{R2}$.

NOTE: Using the indirect method of measuring V_{R1} avoids the possibility of ground loops.

4. Insert the calculated and measured values into Table 13-4.

PART B: Vp-p Voltage

1. Calculate the voltage drops of the circuit of Figure 13-8(a), where the input voltage (V_{in}) of the ac generator is set to 6 Vp-p and 1 kHz.

2. Calculate V_{R1} and V_{R2}.
   a. Calculate V_{R1} where $V_{R1} = V_{in} \times R1/(R1+ R2)$.
   b. Calculate V_{R2} where $V_{R2} = V_{in} \times R2/(R1+ R2)$.

3. Open and select Figure 13-8(a) from the File menu, or connect the circuit. Measure V_{R2}, as shown in Figure 13-8(a). Then, determine the voltage drop across R1 indirectly where: $V_{R1} = V_{in} - V_{R2}$.

NOTE: In practice, one end of the oscilloscope lead is usually grounded, so using the indirect method of measuring V_{R1} avoids the possibility of ground loops.

4. Insert the calculated and measured values into Table 13-4.

Introduction to AC Sine Wave Analysis and Oscilloscope Measurements — 95

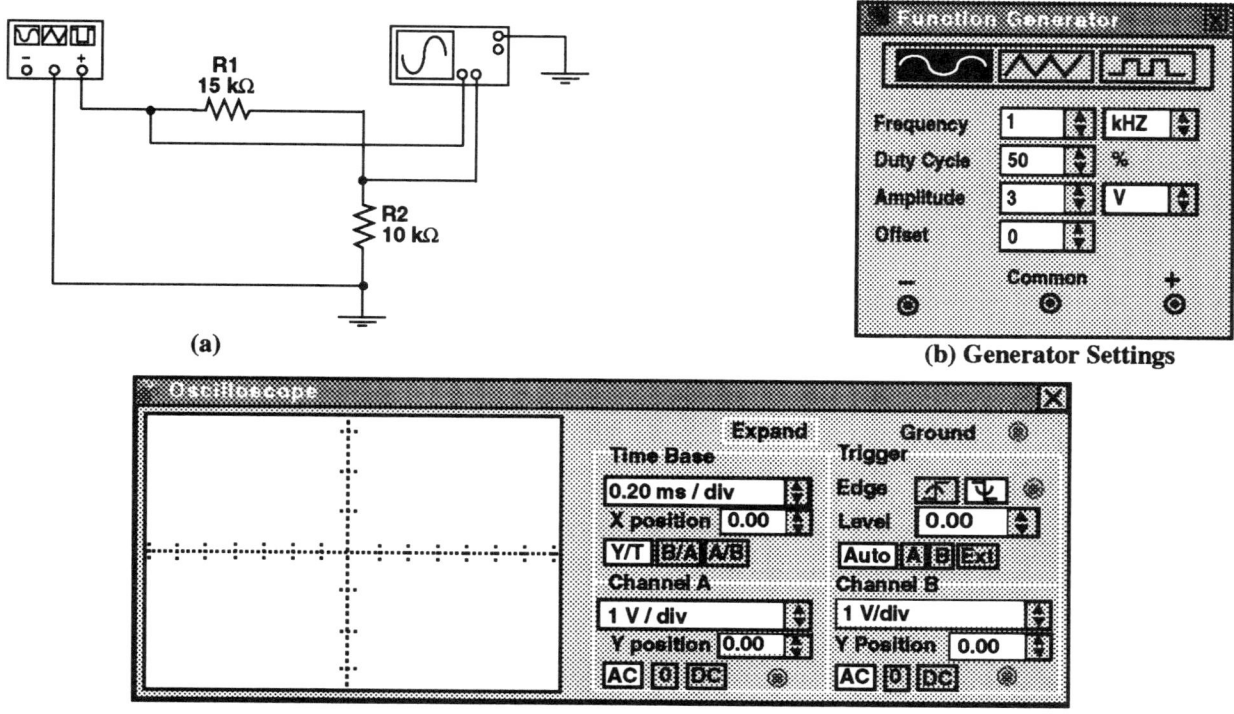

(a)

(b) Generator Settings

(c) Oscilloscope Settings
FIGURE 13-8

TABLE 13-4	Voltmeter rms Voltage			Scope Vp-p Voltage		
	V_{in}	V_{R1}	V_{R2}	V_{in}	V_{R1}	
CALCULATED	V	V	V	Vp-p	Vp-p	Vp-p
MEASURED	V	V	V	Vp-p	Vp-p	Vp-p

SECTION IV: Converting Vp-p and V rms Voltages

PART A: V rms to Vp-p Voltages

1. Select Figure 13-9(a) from the File menu, and open, or connect the ac voltage source and the oscilloscope, as shown in Figure 13-9(a). Initially, the voltage source is set at 1 kHz and 1 V rms. The scope is set a 1 V/div and 0.5 ms/div. Then, 1 V, 2 V, and 3 V of rms voltage are converted to peak-to-peak voltage using Vp-p = $2\sqrt{2}$ V rms.

2. Measure the Vp-p voltage on the oscilloscope where V rms is set to 1 V.

3. Repeat the Vp-p voltage measurements where V rms is set to 2 V and 3 V.

4. Insert the measured Vp-p values, as indicated, into Table 13-5.

PART B: Vp-p to V rms Voltages

1. Select Figure 13-10(a) from the File menu and open, or connect the function generaator and the multimeter as shown in Figure 13-10(a). Initially the function generator is set at 1 kHz and 2 Vp-p (1 Vpeak). Then, 2 Vp-p, 4 Vp-p, and 6 Vp-p will be converted to rms voltage using V rms = Vp-p/$2\sqrt{2}$.

96 — Basic Circuit Analysis For Electronics Using Electronics Workbench®

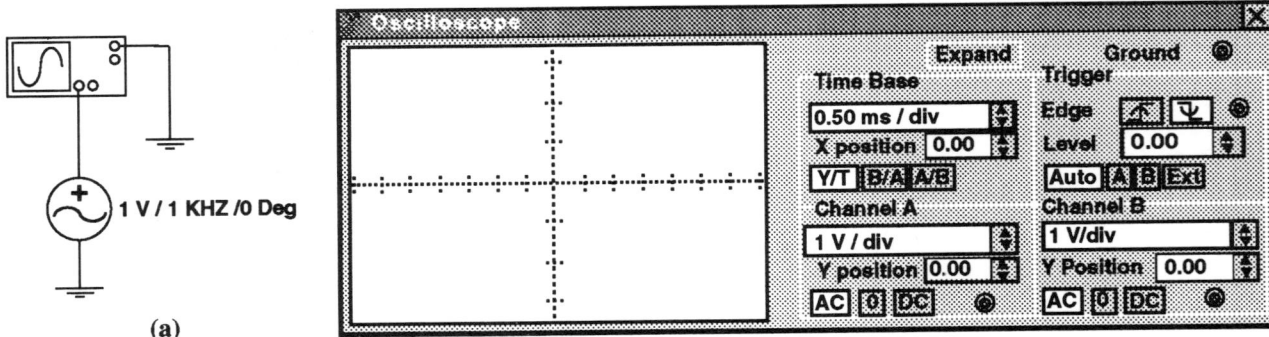

(a)

(b) Oscilloscope Settings

FIGURE 13-9

2. Measure the V rms voltage on the oscilloscope where Vp-p is set to 2 Vp-p.

3. Repeat the V rms voltage measurements where V p-p is set to 4 Vp-p and 6 Vp-p.

4. Insert the calculated and measured V rms values, as indicated, into Table 13-5.

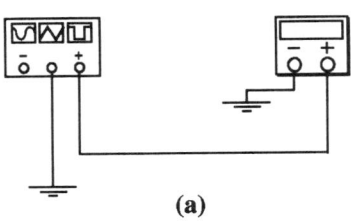

(a)

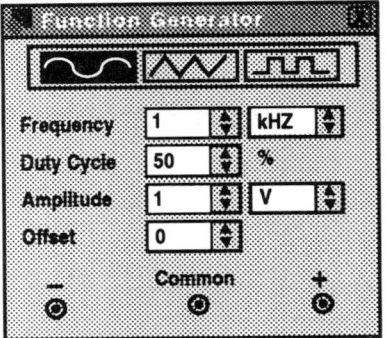

(b) Generator Settings

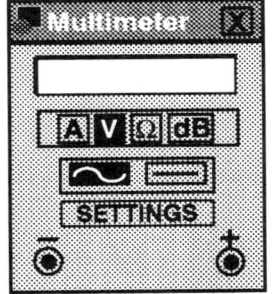

(c) Multimeter Settings

FIGURE 13-10

TABLE 13-5	Converting V rms to Vp-p			Converting Vp-p to V rms		
	1 V	2 V	3 V	2 Vp-p	4 Vp-p	6 Vp-p
CALCULATED	Vp-p	Vp-p	Vp-p	V	V	V
MEASURED	Vp-p	Vp-p	Vp-p	V	V	V

Questions and Problems: Basic Circuit Analysis for Electronics: 198-199

CHAPTER 14
COMBINED DC AND AC CIRCUIT ANALYSIS

INTRODUCTION

Large value electrolytic capacitors are effective "opens" to the dc voltage and effective "shorts" to the ac signal voltage at mid-band frequencies. So, in the analysis of circuits containing both a dc power supply voltage source and an ac signal source, it is easier to do the dc and ac analysis of the circuit separately. A basic three resistor circuit that combines both dc and ac is shown in Figure 14-1. The associated dc circuit formulas used to solve the dc voltage drops of the circuit are provided with the figure. Since the C1 capacitor blocks the flow of dc current through resistor R1, $V_{R1} = 0.0$ V. Also, the associated ac circuit formulas are provided, where V_{in} is passed by C1, since it is an effective short at midband frequencies, which allows the ac signal voltage to be distributed across R1 and the parallel combination of R2 and R3.

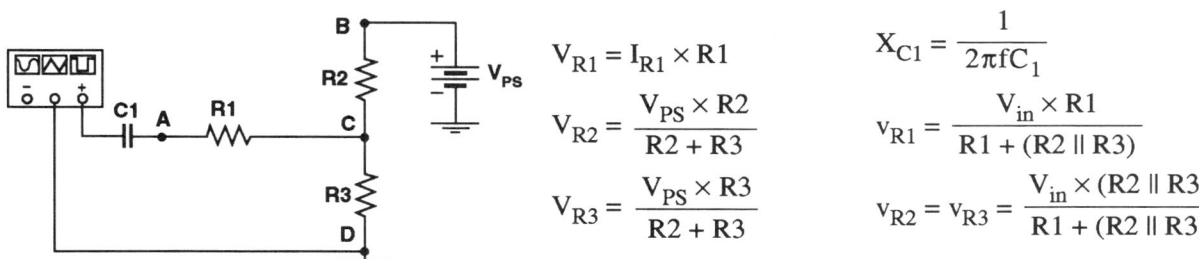

Figure 14-1: Basic circuit.   DC Circuit Formulas   AC Circuit Formulas

$$V_{R1} = I_{R1} \times R1 \qquad X_{C1} = \frac{1}{2\pi f C_1}$$

$$V_{R2} = \frac{V_{PS} \times R2}{R2 + R3} \qquad v_{R1} = \frac{V_{in} \times R1}{R1 + (R2 \parallel R3)}$$

$$V_{R3} = \frac{V_{PS} \times R3}{R2 + R3} \qquad v_{R2} = v_{R3} = \frac{V_{in} \times (R2 \parallel R3)}{R1 + (R2 \parallel R3)}$$

MORE COMPLEX CIRCUIT

In the more complex circuit of Figure 14-2, both C1 and C2 are effective opens to the dc voltage and shorts to the ac signal voltage. The associated dc circuit formulas are provided in conjunction with the circuit, where capacitor C1 blocks the flow of dc current through resistor R1, so $V_{R1} = 0.0$ V. Since it is an effective open to the dc voltage the C2 capacitor connected across the R4 resistor, has no effect on the dc distribution across R4. Also, the associated ac circuit formulas are provided, where V_{in} is passed by C1 to distribute the signal voltage around the circuit. However, bypass capacitor C2 is an effective short at the midband frequencies and, since it is in parallel with resistor R4, no ac signal voltage is dropped across R4.

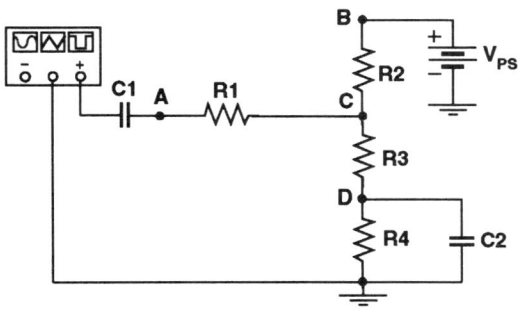

Figure 14-2: Basic circuit.   DC Circuit Formulas   AC Circuit Formulas

$$V_{R1} = I_{R1} \times R1 \qquad X_{C1} = \frac{1}{2\pi f C_1}, \ X_{C2} = \frac{1}{2\pi f C_2}$$

$$V_{R2} = \frac{V_{PS} \times R2}{R2 + R3 + R4} \qquad v_{R1} = \frac{V_{in} \times R1}{R1 + (R2 \parallel R3)}$$

$$V_{R3} = \frac{V_{PS} \times R3}{R2 + R3 + R4} \qquad v_{R2} = v_{R3} = \frac{V_{in} \times (R2 \parallel R3)}{R1 + (R2 \parallel R3)}$$

$$V_{R4} = \frac{V_{PS} \times R4}{R2 + R3 + R4} \qquad v_{R4} = 0.0 \text{ V}$$

LABORATORY EXERCISE

READING ASSIGNMENT: Basic Circuit Analysis for Electronics: 200-206.

EXERCISE OBJECTIVES

To become familiar with:

- The analysis of circuits that contain both dc voltage supplies and ac signal sources.

- Large valued electrolytic capacitors that are effectively "opens" to dc voltages and effectively "shorts" to ac signal voltages.

- How electrolytic capacitors allow dc voltages and ac signal voltages to be solved separately, and how both terminals of the dc power supply are effectively "ac grounds" to the ac signal.

PROCEDURE

SECTION I: Basic three Resistor Circuit

PART A: Pre-laboratory Calculations

1. For the circuit of Figure 14-3, the dc Vps voltage is 18 V and the input ac signal voltage is 6 Vp-p.

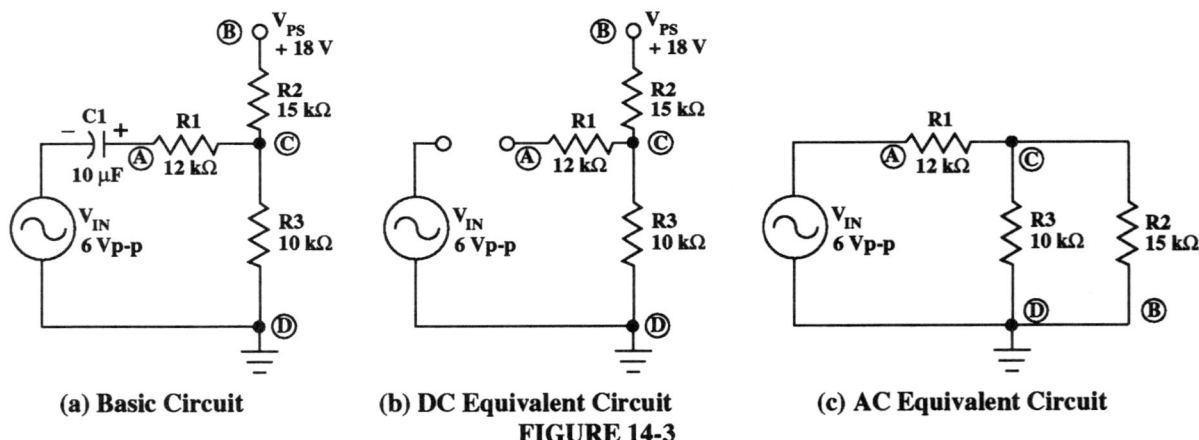

(a) Basic Circuit (b) DC Equivalent Circuit (c) AC Equivalent Circuit

FIGURE 14-3

2. DC Calculations:

   a. Calculate the dc voltage drops of the circuit where, as shown, in the dc equivalent circuit of Figure 14-3(b), the C1 coupling capacitor looks like an "effective open" to the dc voltage.

   b. Calculate V_{R2}, the dc voltage drop across R2, where: $V_{R2} = (V_{PS} \times R2)/(R2 + R3)$.

   c. Calculate V_{R3}, the dc voltage drop across R3, where: $V_{R3} = (V_{PS} \times R3)/(R2 + R3)$.

   d. Calculate V_{R1}, the dc voltage drop across R1, where: $V_{R1} = I_{R1} \times R1$, where $I_{R1} \approx 0.0$ mA.

NOTE: Since R1 is effectively "open circuited" by coupling capacitor C1, no current should flow through resistor R1 if the correct polarity of the electrolytic capacitor is observed.

Combined DC and AC Circuit Analysis — 99

3. AC Calculations:
    a. Calculate the ac signal voltage drops of the circuit where, as shown in the ac equivalent circuit of Figure 14-3(c), V_{in} = 6 Vp-p and 1 kHz, and the C1 coupling capacitor looks like an "effective short" to the ac signal voltage.
    b. Calculate v_{R1}, the ac signal voltage drop across R1, where $v_{R1} = (V_{in} \times R1)/(R1 + [R2 \| R3])$.
    c. Calculate $v_{(R2\|R3)}$, the ac signal voltage drop across R2 ‖ R3, where:
    $$v_{(R2\| R3)} = (V_{in} \times [R2 \| R3])/(R1 +[R2 \| R3]).$$

NOTE: At mid-band frequencies, typically 1 kHz, large value electrolytic capacitors are effective "short circuits" and no ac voltage is dropped across the capacitor.

PART B: DC Voltage Measurements

Open and select from the circuit of Figure 14-4(a) from the File menu, or connect the circuit. The multimeter is set to dc (—) and volts (V) as shown in Figure 14-4(b).

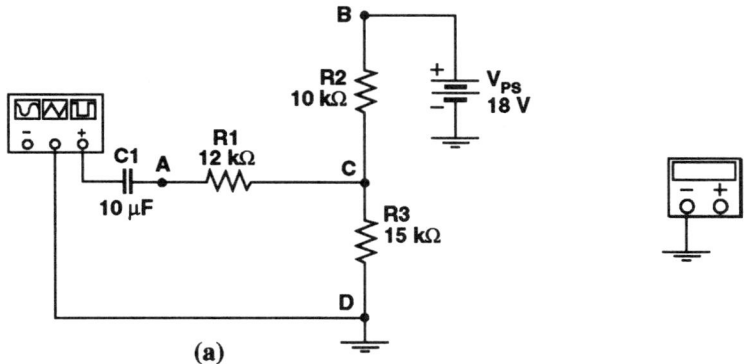
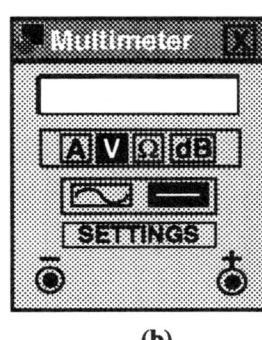

(a)   (b)

FIGURE 14-4

1. DC Voltage Measurements:
    a. Measure V_A, V_B, and V_C, the single point dc voltages of the circuit, all measured with respect to point D, the reference ground of the circuit.
    b. Then, use the measured V_A, V_B, and V_C values to obtain the "measured" V_{R1}, V_{R2}, and V_{R3} values, where $V_{R1} = V_A - V_C$, $V_{R2} = V_B - V_C$, and $V_{R3} = V_C - V_D$.

NOTE: Measuring the dc voltage indirectly, with reference to ground, avoids possible ground loops.

2. Insert the calculated and measured dc voltage values, as indicated, into Table 14-1.

TABLE 14-1	V_{R1}	V_{R2}	V_{R3}	V_A	V_B	V_C
CALCULATED						
MEASURED						

PART C: AC Voltage Measurements

Open and select the circuit of Figure 14-5(a) from the File menu, or connect the circuit. Use an oscilloscope to measure the circuit ac. Set the generator to 1 kHz and 6 Vp-p, as shown in Figure 14-5(b). Make sure the oscilloscope is set to 1 V/div and 0.5 ms/div as shown in Figure 14-5(c)

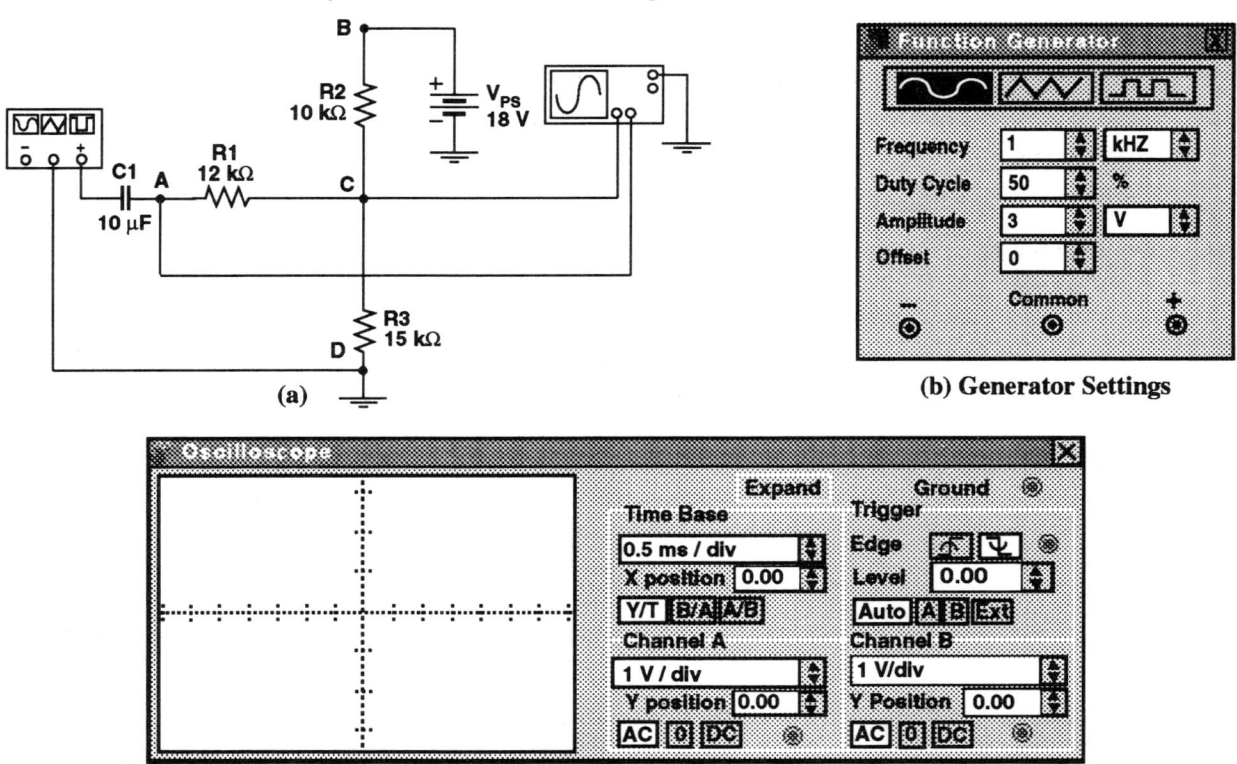

(a)

(b) Generator Settings

(c) Oscilloscope Settings
FIGURE 14-5

1. AC Voltage Measurements:
   a. Measure V_{in} and v_C, the single point ac voltages of the circuit, both measured with respect to point D, the reference ground of the circuit.
   b. Then, use the measured V_{in} at 6 Vp-p and v_C values to obtain the "measured" v_{R1} and $v_{R2} = v_{R3}$ values where: $v_{R1} = v_A - v_C$ and $v_{R2} = v_{R3} = v_C$.

2. Next, monitor the ac signal voltage riding on the dc. Begin by setting the volts/div at 5 V/div and AC mode, and keep the time/div set at 0.5 ms/div. Then, switch to the DC mode, noting the dc voltage shift and the ac signal riding on the dc voltage.

NOTE: Measuring the ac voltage indirectly avoids the possibility of ground loops. Also, because the dc voltage is high with regard to the ac signal voltage, it is necessary to increase the volt/div to 5 V/div in order to see the ac riding on the dc voltage.

3. Insert the calculated and measured ac signal voltage values, as indicated, into Table 14-2.

TABLE 14-2	$V_{in} = v_A$	$v_C = v_{R2} = v_{R3}$	v_{R1}
CALCULATED	6 Vp-p		
MEASURED			

SECTION II: More Complex Four Resistor Circuit
PART A: Pre-laboratory Calculations:
For the circuit of Figure 14-6 the dc Vps is 18 V and the ac input voltage is 6 Vp-p at 1 kHz.

Combined DC and AC Circuit Analysis — 101

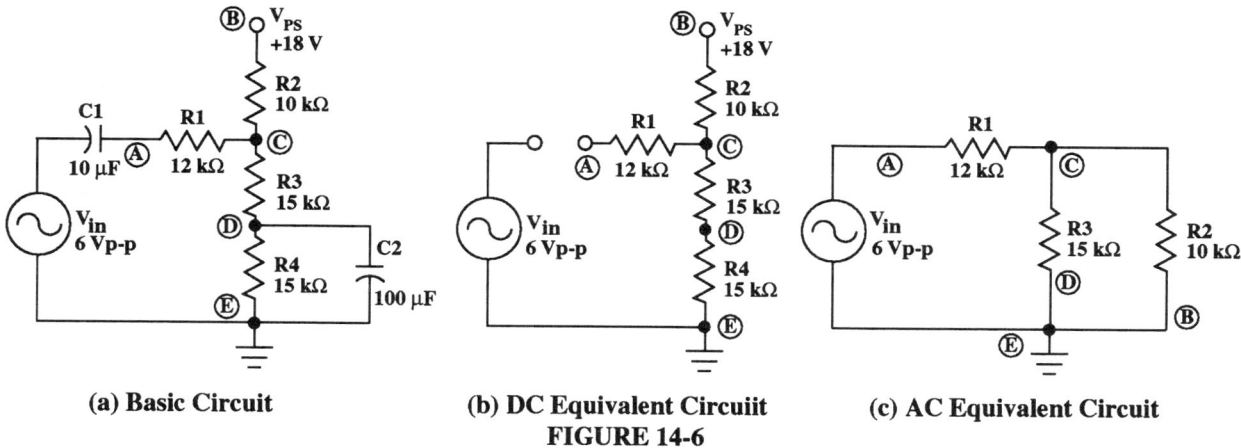

(a) Basic Circuit   (b) DC Equivalent Circuiit   (c) AC Equivalent Circuit

FIGURE 14-6

1. DC Calculations:
   a. Calculate the dc voltage drops of the circuit where, as shown in the dc equivalent circuit of Figure 14-6(b), the C1 coupling capacitor looks like an "effective open" to the dc voltage.
   b. Calculate V_{R2}, the dc voltage drop across R2, where: $V_{R2} = (V_{PS} \times R2)/(R2 + R3 + R4)$.
   c. Calculate V_{R3}, the dc voltage drop across R3, where: $V_{R3} = (V_{PS} \times R3)/(R2 + R3 + R4)$.
   d. Calculate V_{R4}, the dc voltage drop across R4, where: $V_{R4} = (V_{PS} \times R4)/(R2 + R3 + R4)$.
   e. Calculate V_{R1}, the dc voltage drop across R1, where: $V_{R1} = I_{R1} \times R1$ and $I_{R1} \approx 0.0$ mA.

NOTE: Since C1 and C2 are "open circuits" to dc voltage, no dc current should flow through them. But C1 and C2 are effective shorts to ac signal voltage, so C2 causes the junction of R3 and R4 to be at ac ground.

2. AC Calculations:
   a. Calculate the ac signal voltage drops of the circuit where, as shown in the ac equivalent circuit of Figure 14-6(c), V_{in} = 6 Vp-p and 1 kHz and the C1 and C2 capacitors looks like "effective shorts" to the ac signal voltage.
   b. Calculate v_{R1}, the ac signal voltage drop across R1, where: $v_{R1} = (V_{in} \times R1)/(R1 + [R2 \| R3])$.
   c. Calculate $v_{R2} = v_{R3}$, the ac signal voltage drop across R2 ∥ R3, where:
   $v_{R2 \| R3} = (V_{in} \times [R2 \| R3])/(R1 + [R2 \| R3])$.

NOTE: At mid-band frequencies, typically 1 kHz, no signal voltage is dropped across either C1 or C2.

PART B: Measurements

1. DC Voltage Measurements: Open and select from the File menu the circuit of Figure 14-7(a), or connect the circuit. The multimeter is set to dc (—) and volts (V), as shown in Figure 14-7(b).
   a. Measure V_A, V_B, V_C, and V_D, the single point dc voltages of the circuit, all measured with respect to point E, the reference ground of the circuit.
   b. Then, use the measured V_A, V_B, V_C, and V_D values to obtain the "measured" V_{R1}, V_{R2}, V_{R3}, and V_{R4} values.

NOTE: Measuring the dc voltage indirectly avoids possible ground loops.

2. Insert the calculated and measured dc voltage values, as indicated, into Table 14-3.

102 — Basic Circuit Analysis For Electronics Using Electronics Workbench®

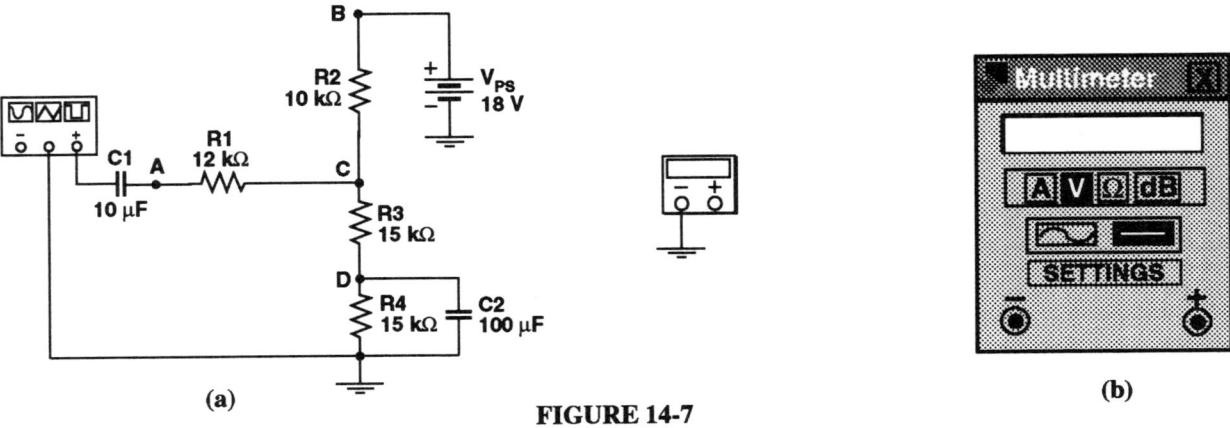

FIGURE 14-7

TABLE 14-3	V_{R1}	V_{R2}	V_{R3}	V_{R4}	V_A	V_B	V_C	V_D
CALCULATED								
MEASURED								

3. AC Voltage Measurements: Open and select, or connect, the circuit of Figure 14-8. Set the oscilloscope to 1 V/div and 0.5 ms/div, and set the generator to 1 kHz and 3 V so that V_{in} equals 6 Vp-p.
   a. Measure V_{in} and v_C, the single point ac voltage of the circuit, both measured with respect to point E, the reference ground of the circuit, as shown in Figure 14-8(a).
   b. Then, use the measured V_{in} at 6 Vp-p and v_C values to obtain the "measured" v_{R1} and $v_{R2} = v_{R3}$ values.

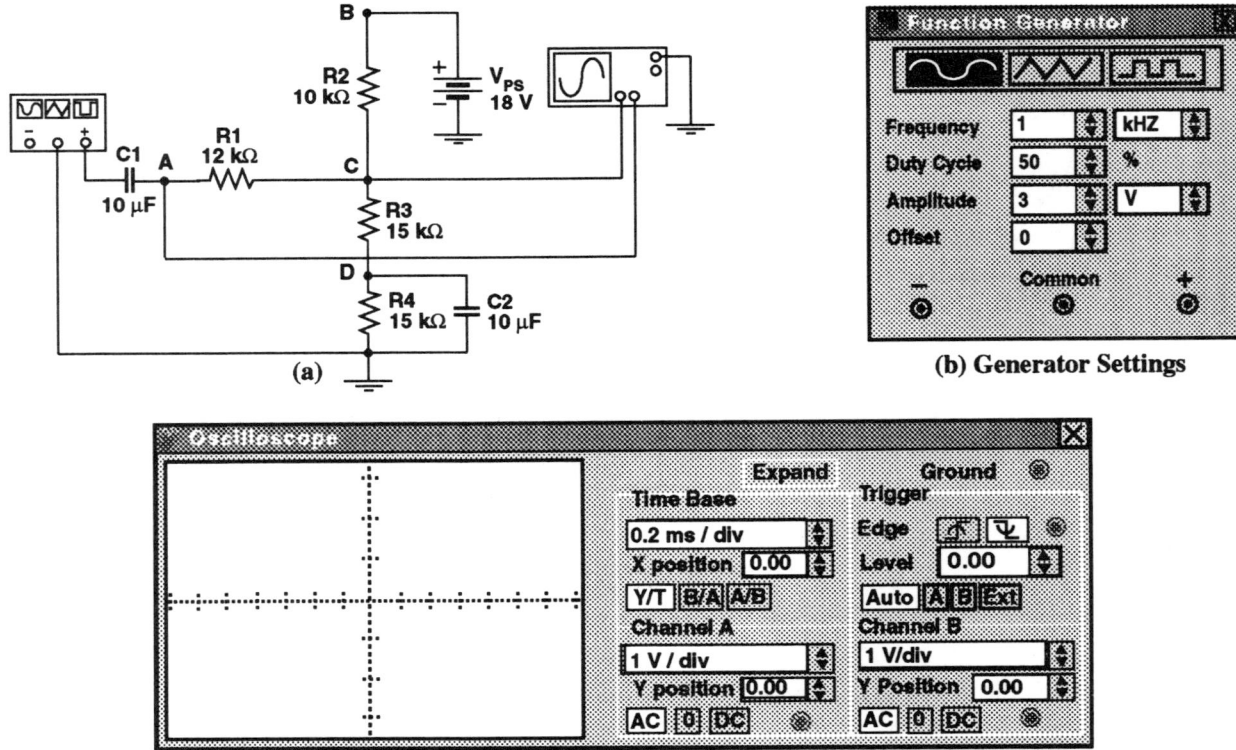

(c) Oscilloscope Settings

FIGURE 14-8

4. Next, monitor the ac signal voltage riding on the dc. Begin by setting the volts/div at 5 V/div and AC mode, and keep the time/div set at 0.5 ms/div. Then, switch to the DC mode, noting the dc voltage shift and the ac signal riding on the dc voltage.

NOTE: Measuring the ac voltage indirectly avoids the possibility of ground loops. Also, because the dc voltage is high with regard to the ac signal voltage, it is necessary to increase the volt/div to 5 V/div in order to see the ac riding on the dc voltage.

5. Insert the calculated and measured ac signal voltage values, as indicated, into Table 14-4.

TABLE 14-4	$V_{in} = v_A$	$v_C = v_{R2} = v_{R3}$	v_D	v_{R1}
CALCULATED	6 Vp-p			
MEASURED				

Questions and Problems: Basic Circuit Analysis for Electronics: 211-212

CHAPTER 15
CAPACITORS IN AC SINE WAVE CIRCUITS

INTRODUCTION

Capacitors are the second most widely used passive components in electronic circuits after resistors. Capacitors along with resistors and inductors are used in filter circuits, that pass or reject ac signal voltages at selected frequencies. The formula for determining the capacitive reactance of capacitors is $X_C = 1/2\pi fC$, where f is the frequency in Hertz and C is the capacitance in Farads. Therefore, in two capacitor series circuit as shown in Figure 15-1(a), X_{C1} and X_{C2} are solved to find X_{CT}. Then, Ohm's law is used to solve I and, in turn, the voltage drops across the C1 and C2 capacitors.

In the series resistor and capacitor circuit of Figure 15-1(b), X_C is solved and the impedance Z of the circuit. Then, Ohm's law is then used to solve the circuit current I and the voltage drops across the C and R circuit components.

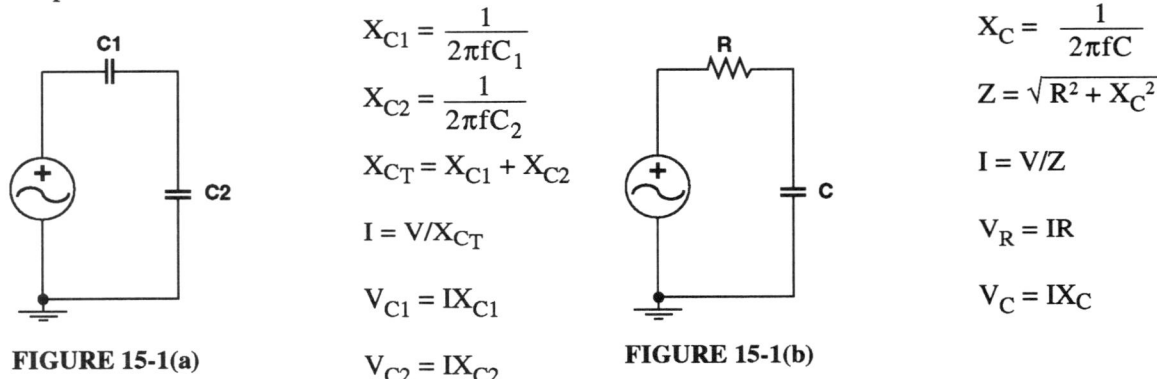

FIGURE 15-1(a) **FIGURE 15-1(b)**

In the resistor and two series capacitor of Figure 15-2(a), C_T is found from $(C1 \times C2)/(C1 + C2)$ and once X_{CT} is calculated, and impedance Z of the circuit is solved. Ohms law is used to solve the voltage drops across R and across the series C1 and C2 capacitors. In Figure 15-2(b), C_T is solved by adding C1 and C2, X_{CT} is solved, and then the impedance Z of the circuit solved. Ohms law is then used to solve I, and then the voltage drops across the R resistor and across the parallel C1 and C2 capacitors.

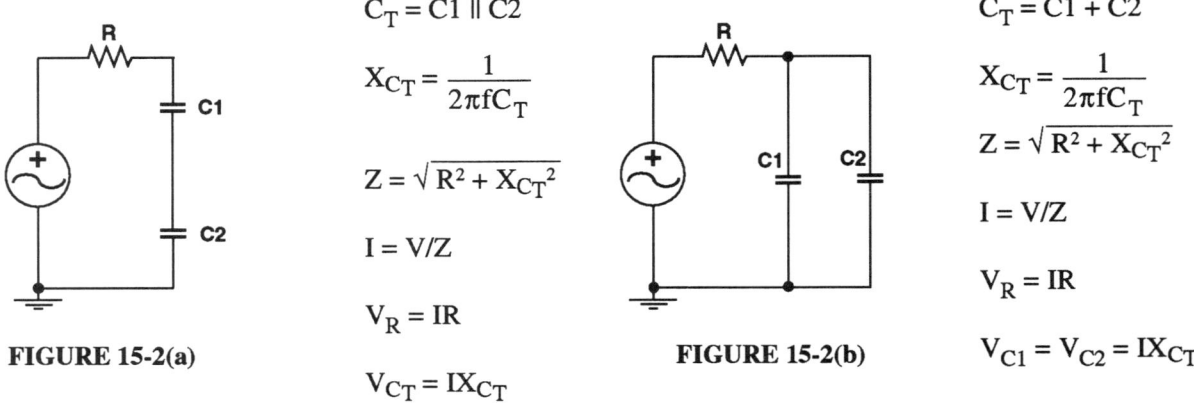

FIGURE 15-2(a) **FIGURE 15-2(b)**

Capacitors in AC Sine Wave Circuits — 105

LABORATORY EXERCISE

READING ASSIGNMENT: Basic Circuit Analysis for Electronics: 213-220.

EXERCISE OBJECTIVES
To become familiar with:

- ac voltage divider circuits using series capacitors.
- ac voltage divider circuits using series and parallel R and C components.
- Phase shift measurements.

PROCEDURE

SECTION I: Capacitor Voltage Divider Circuits
PART A: Equal Value C1 and C2 Capacitors
Pre-laboratory Calculations

1. Analyze the two capacitors series circuit of Figure 15-3(a), where C1 and C2 are each 0.01µF.

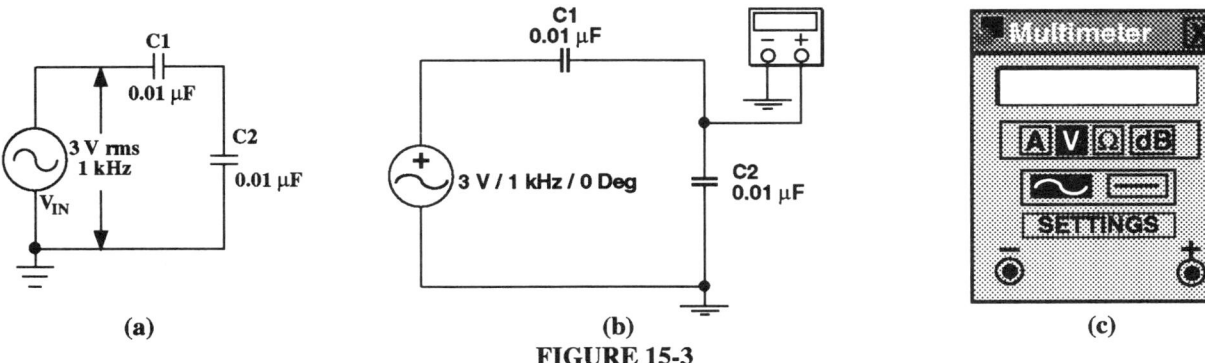

FIGURE 15-3

2. Calculate the output voltage drop across capacitors C1 and C2, where V_{in} of the generator is set at 3 V rms and 1 kHz.

   a. Solve the capacitive reactance of the C1 = C2 = 0.01µF capacitors, where:

   $X_{C1} = 1/2\pi f C_1$ and $X_{C2} = 1/2\pi f C_2$.

   b. Calculate the circuit current I, where $I = V_{in}/(X_{C1} + X_{C2})$.

   c. Calculate the ac voltage drop across the C1 capacitor, where: $V_{C1} = I(X_{C1})$.

   d. Calculate the ac voltage drop across the C2 capacitor, where: $V_{C2} = I(X_{C2})$.

3. Insert the calculated values, as indicated, into Table 15-1.

Voltmeter Circuit Measurements

Open and select from the File menu the circuit diagram of Figure 15-3(b), or connect the circuit. The multimeter is set to ac (~) and volts (V), as shown in Figure 15-3(b). In the circuit the V_{in} of the generator is set to 3 V rms and the frequency to 1 kHz.

1. Measure V_{C2}, the ac voltage drops across the C2 capacitor.

2. Then find V_{C1}, the ac voltage drop across the C1 capacitor, indirectly, where: $V_{C1} = V_{in} - V_{C2}$.

PART B: Unequal Value C1 and C2 Capacitors
Pre-Lab Calculations
1. Analyze the two capacitors series circuit of Figure 15-4(a) where: C1 = 0.01 µF and C2 = 0.02 µF.

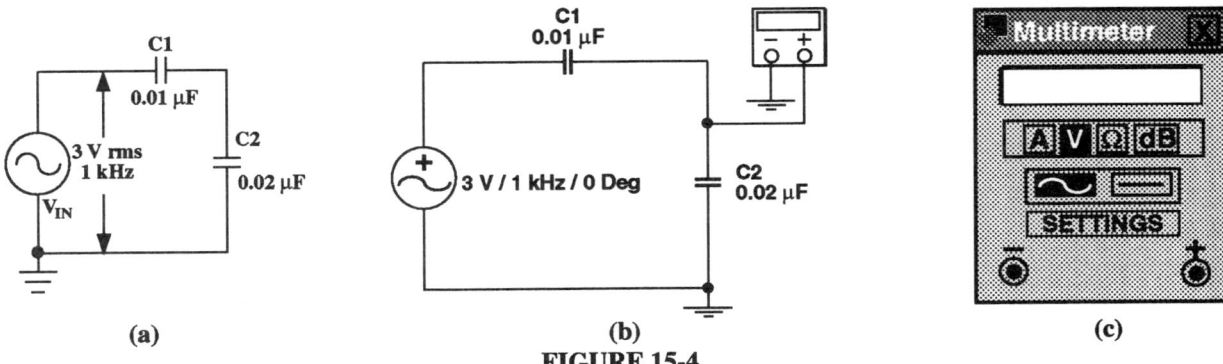

(a)   (b)   (c)

FIGURE 15-4

2. Calculate the output voltage drop across capacitor C1 and C2, where V_{in} of the generator is set at 3 V rms and the frequency to 1 kHz.
   a. Solve the capacitive reactance of the C1 = 0.01 µF and C2 = 0.02 µF capacitors, where:
      $X_{C1} = 1/2\pi f C_1$ and $X_{C2} = 1/2\pi f C_2$.
   b. Calculate the circuit current I, where $I = V_{in}/(X_{C1} + X_{C2})$.
   c. Calculate the ac voltage drop across the C1 capacitor, where $V_{C1} = I(X_{C1})$.
   d. Calculate V_o, the ac voltage drop across the C2 capacitor, where $V_{C2} = I(X_{C2})$.

3. Insert the calculated values, as indicated, into Table 15-1.

Voltmeter Circuit Measurements
1. Open and select from the File menu the circuit diagram of Figure 15-4(b), or connect the circuit. The multimeter is set to ac (~) and volts (V). The V_{in} of the generator is set to 3 V rms and the frequency to 1 kHz.
2. Measure V_{C2}, the ac voltage drops across the C2 capacitor.
3. Then find V_{C1}, the ac voltage drop across C1 capacitor indirectly, where $V_{C1} = V_{in} - V_{C2}$.
4. Insert the measured values, as indicated, into Table 15-1.

TABLE 15-1	Part A: C1 = C2					Part B: C1 ≠ C2				
	X_{C1}	X_{C2}	I	V_{C1}	V_{C2}	X_{C1}	X_{C2}	I	V_{C1}	V_{C2}
CALC.										
MEAS.										

Section II: Resistor Capacitor circuits
PART A: Resistor-Capacitor Series Circuit.
Pre-laboratory Calculations
1. Analyze the RC series circuit of Figure 15-5(a), where R = 12 kΩ and C = 0.01 µF.
2. Calculate the output voltage drop across capacitor C, where V_{in} of the generator is set at 3 V rms and 1 kHz.
   a. Solve the capacitive reactance of the C = 0.01 µF capacitor, where $X_C = 1/2\pi f C$.
   b. Calculate the impedance Z of the R and X_C series circuit where: $Z = \sqrt{R^2 + X_C^2}$.

Capacitors in AC Sine Wave Circuits — 107

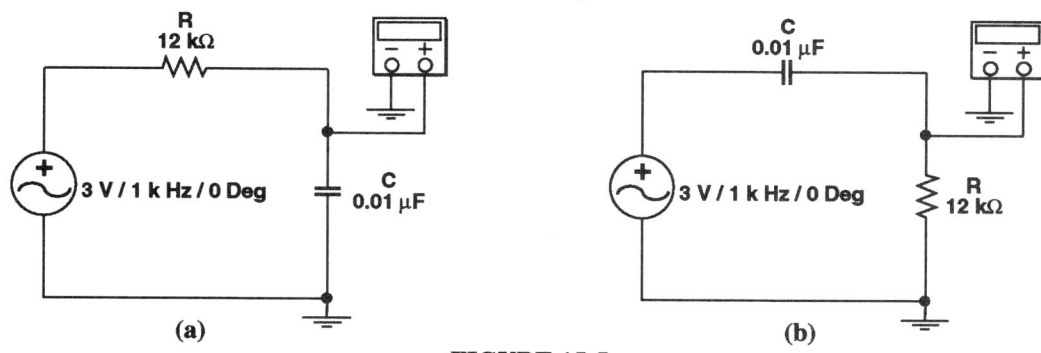

FIGURE 15-5

   c. Calculate the circuit current I, where $I = V_{in}/Z$.
   d. Calculate the ac voltage drop across the resistor R, where $V_R = IR$.
   e. Calculate the ac voltage across the capacitor C, where $V_C = IX_C$.

3. Insert the calculated values, as indicated, into Table 15-2.

Voltmeter Circuit Measurements

Open and select from the File menu the circuit diagram of Figure 15-5(a), or connect the circuit. The multimeter is set to AC (~) and volts (V). The V_{in} of the generator is set to 3 V rms and the frequency to 1 kHz.

1. Measure V_C, the ac voltage drop across the capacitor C.

2. Interchange the resistor and capacitor, as shown in Figure 15-5(b), and maintain V_{in} of the generator at 3 V rms and the frequency at 1 kHz. Measure the ac voltage drops across the 12 kΩ resistor R.

3. Use the measured V_R and V_C to provide the X_C and Z values.
   a. Use the measured V_R and the R value to determine the circuit current, where $I = V_R/R$.
   b. Use the measured V_C and the circuit current I to determine X_C, where $X_C = V_C/I$.
   c. Use V_{in} and the circuit current I to determine Z, where $Z = V_{in}/I$.

4. Insert the measured values, as indicated, into Table 15-2.

TABLE 15-2	X_{C1}	Z	I	V_R	V_C
CALC.					
MEAS.					

PART B: Series R and Parallel C1 and C2 Circuit
Pre-Laboratory Calculations
1. Analyze the RC circuit of Figure 15-6(a), where R = 12 kΩ and C1 = C2 = 0.01 µF.

2. Calculate the output voltage drop across the parallel C1 and C2 capacitors, where V_{in} of the generator is set at 3 V rms and 1 kHz.
   a. Solve C_T where: $C_T = C1 + C2$ and C1 = C2 = 0.01 µF.
   b. Solve the capacitive reactance of C_T where: $X_{CT} = 1/2\pi fC_T$.
   c. Calculate the impedance of R and X_{CT} circuit where: $Z = \sqrt{R^2 + X_{CT}^2}$.
   d. Calculate the circuit current I where: $I = V_{in}/Z$.
   e. Calculate the ac voltage across C_T, the parallel C1 and C2 capacitors, where: $V_{CT} = I(X_{CT})$.
   f. Calculate the ac voltage across resistor R where: $V_R = IR$.

3. Insert the calculated values, as indicated, into Table 15-3.

Voltmeter Circuit Measurements

Open and select the circuit diagram of Figure 15-6(a) from the File menu, or connect the circuit. Set the generator at 3 V rms and the frequency to 1 kHz. The voltmeter is set to ac ($\sim$) and volts (V).

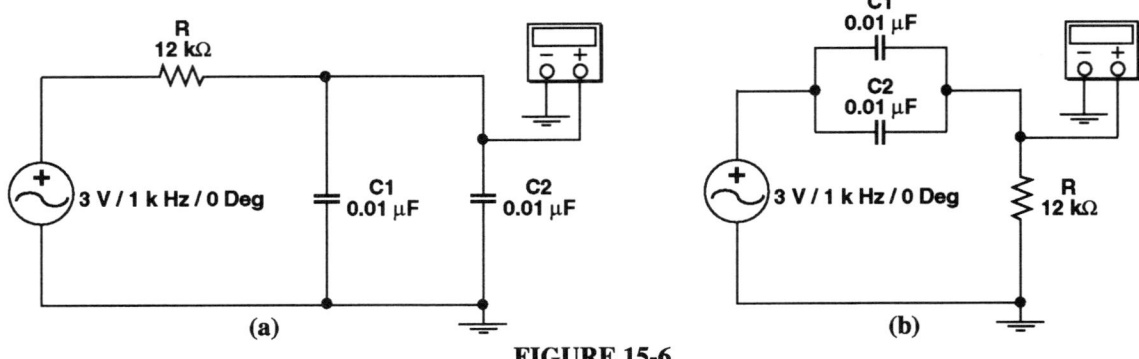

FIGURE 15-6

1. Measure V_{CT}, the ac voltage drop across the parallel C1 and C2 capacitors.

2. Interchange the resistor and parallel capacitors, as shown in Figure 15-6(b). Maintain V_{in} of the generator at 3 V rms and the frequency at 1 kHz. Measure the ac voltage drops across the resistor R.

3. Use the measured V_R and V_{CT} to prove X_{CT} and C_T.
   a. Use the measured V_R and the R values to determine the circuit current where: $I = V_R/R$.
   b. Use the measured V_{CT} and the circuit current to determine X_{CT} where: $X_{CT} = V_{CT}/I$.
   c. Use X_{CT} to determine C_T where: $C_T = 1/2\pi f(X_{CT})$.
   d. Use V_{in} and the circuit current I to determine Z where: $Z = V_{in}/I$.

4. Insert the measured values, as indicated, into Table 15-3.

TABLE 15-3	C_T	X_{CT}	Z	I	V_{CT}	V_R
CALCULATE						
MEASURE						

PART C: Series R1, C1, and C2 Circuit
Pre-laboratory Calculations

1. Analyze the circuit of Figure 15-7(a), where R = 12 kΩ and C1 = C2 = 0.01 µF.

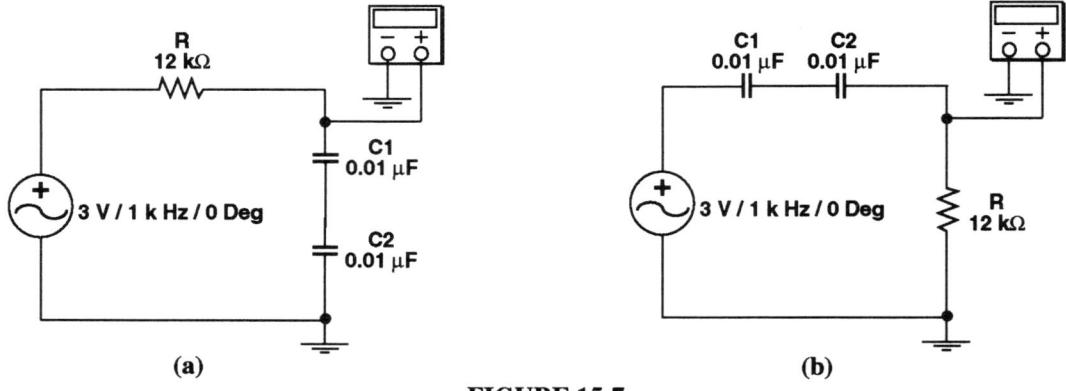

FIGURE 15-7

2. Calculate the output voltage drop across series C1 and C2, where V_{in} of the generator is set at 3 V rms and 1 kHz.
   a. Solve C_T, where $C_T = 1/(1/C1 + 1/C2)$ where: $C1 = C2 = 0.01$ µF.
   b. Solve the capacitive reactance of the C_T where: $X_{CT} = 1/2\pi f C_T$.
   c. Calculate the impedance of the R and X_{CT} series circuit, where $Z = \sqrt{R^2 + X_{CT}^2}$.
   d. Calculate the circuit current I, where $I = V_{in}/Z$.
   e. Calculate the ac voltage across C_T, the series C1 and C2 capacitors, where: $V_{CT} = I(X_{CT})$.
   f. Calculate the ac voltage across resistor R where: $V_R = IR$.

3. Insert the calculated values, as indicated, into Table 15-4.

Voltmeter Circuit Measurements

Open and select the circuit diagram of Figure 15-7(a) from the File menu, or connect the circuit. Set the multimeter to ac (~) and volts (V). The generator is set to 3 V rms at 1 kHz.
1. Measure V_{CT}, the ac voltage drop across the series C1 and C2 capacitors.

2. Interchange the resistor and series capacitors in the circuit, as shown in Figure 15-7(b). Maintain V_{in} of the generator at 3 V rms and the frequency at 1 kHz. Measure the ac voltage drops across resistor R.

3. Use the measured V_R and V_{CT} to provide the X_{CT} and C_T values.
   a. Use the measured V_R and the R values to determine the circuit current where: $I = V_R/R$.
   b. Use the measured V_{CT} and the circuit current to determine X_{CT} where: $X_{CT} = V_{CT}/I$.
   c. Use X_{CT} to determine series C_1 and C_2 capacitors where: $C_T = 1/2\pi f(X_{CT})$.
   d. Use V_{in} and the circuit current I to determine Z where: $Z = V_{in}/I$.

4. Insert the measured values, as indicated, into Table 15-4.

TABLE 15-4	C_T	X_{CT}	Z	I	V_{CT}	V_R
CALCULATE						
MEASURE						

SECTION III: Oscilloscope Phase Shift Measurements
PART A: Pre-laboratory Calculations
1. Analyze the RC series circuit of Figure 15-8(a), where $f = 1/2\pi RC$, $R = 12$ kΩ, and $C = 0.01$ µF.

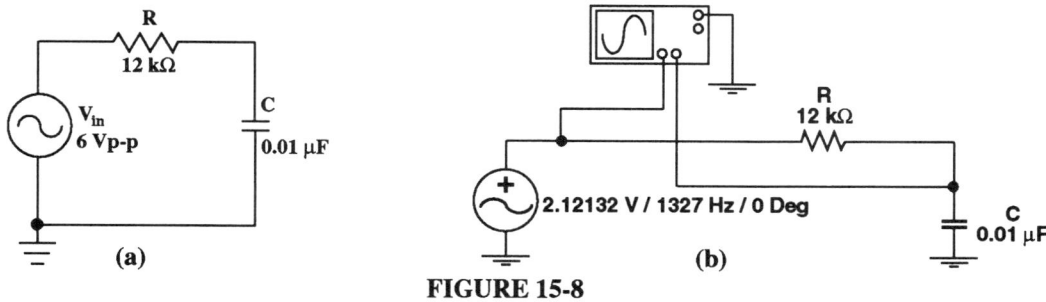

FIGURE 15-8

2. Calculate the output voltage drop across capacitor C, where V_{in} of the generator is set at 6 Vp-p and the operating frequency where: $X_C = R$.
   a. Calculate the operating frequency where: $f = 1/2\pi RC$.

110 — Basic Circuit Analysis For Electronics Using Electronics Workbench®

  b. Calculate the capacitive reactance of the C = 0.01 µF capacitor where: $X_C = 1/2\pi fC$.
  c. Calculate the impedance Z, of the R and X_C series circuit where: $Z = \sqrt{R^2 + X_C^2}$.
  d. Calculate the phase angle $\angle\theta$ where: $\angle\theta = \tan^{-1} (-) X_C/R$.
  e. Calculate the circuit current I where: $I = V_{in}/Z\angle\theta$.
  f. Calculate the ac voltage drop across resistor R where: $V_R = I\angle\theta \times (R)$.
  g. Calculate the ac voltage drop across capacitor C where: $V_C = I\angle\theta \times (X_C\angle\theta)$.

3. Insert the calculated values, as indicated, into Table 15-5.

PART B: Scope Circuit Measurements of Figure 15-8

Open and select the circuit diagram of Figure 15-9(b) from the File menu, or connect the circuit. Set the generator to 6 Vp-p at the calculated frequency. The oscilloscope is set to the ac mode: 1 V/div – 0.1 ms/div.

1. Amplitude Measurements: Measure the amplitude of the input voltage V_{in} and the output voltage V_o across the C capacitor.
   a. Measure the amplitude of the input voltage V_{in}. Click on the scope to enlarge it, and click on the power switch to activate the circuit. Select Expand and use cursor #2 to monitor the positive-going wave and use cursor #1 to monitor the negative-going wave. Read the peak-to-peak voltage swing directly from: $V_{in} = VA2 - VA1$.
   b. Measure the amplitude of the smaller in amplitude output voltage V_o. Use cursor #2 to monitor the positive-going wave of the phase-lagging wave (≈ 0.707 of V_{in}) and cursor #1 to monitor the negative-going wave. Read the peak-to-peak voltage directly.
2. Phase Measurements: Measure the time of one cycle and measure the time difference between V_{in} and V_o.
   a. Measure the time of one cycle by using cursor #1 and cursor #2 to monitor between the positive-going peaks of one complete cycle and read the time directly from T2 – T1.
   b. Measure the time between the input and output waves (time differential). Set cursor #1 at the positive-going peak of V_{in} and cursor #2 at the positive-going peak of the phase lagging wave. Read the time differential between the two waves and read the time directly where the time differential = T2 – T1.
   c. Determine the phase angle, where $\angle\theta = \dfrac{\text{time difference} \times 360°}{\text{time of one cycle}}$.

NOTE: If cursors #1 and #2 are interchanged, the results will be indicated by a minus (–) sign. It is also possible to measure the time of a complete cycle and the phase angle between the input and output waves by measuring between the 0.0 crossover points.

3. Insert the measured values, as indicated, into Table 15-6. Then, sketch the amplitude and phase shift of the V_{in} and V_o waveshapes on the graph of Figure 15-10(a) and label.

PART C: Scope Circuit Measurements of Figure 15-9

Open and select the circuit diagram of Figure 15-9(b) from the File menu, or connect the circuit. Set the generator to 6 Vp-p at the calculated frequency. The oscilloscope is set to the ac mode: 1 V/div - 0.1 ms/div.

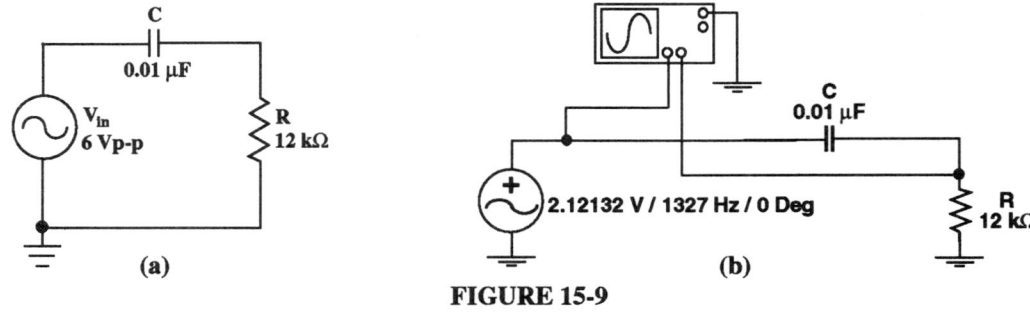

FIGURE 15-9

1. Amplitude Measurements: Measure the amplitude of the input voltage V_{in} and the output voltage V_o across the R resistor.

Capacitors in AC Sine Wave Circuits

a. Measure the amplitude of the input voltage V_{in}. Click on the scope to enlarge it, and click on power switch to activate the circuit. Select Expand and use cursor #2 to monitor the positive-going wave and use cursor #1 to monitor the negative-going wave. Read the peak-to-peak voltage swing directly from: $V_{in} = VA2 - VA1$.

b. Measure the amplitude of the smaller in amplitude output voltage V_o. Use cursor #2 to monitor the positive-going wave of the phase-lagging wave (≈ 0.707 of V_{in}) and use cursor #2 to monitor the negative-going wave. Read the peak-to-peak voltage swing directly.

2. Phase Measurements: Measure the time of one cycle and measure the time difference between V_{in} and V_o.

    a. Measure the time of one cycle by using cursor #1 and cursor #2 to monitor between the positive-going peaks of one complete cycle. Read the time directly from T2 – T1.

    b. Measure the time between the input and output waves (time differential). Set cursor #1 at the positive-going peak of V_{in} and cursor #2 at the positive-going peak of the phase leading wave. Read the time differential between the two waves and read the time directly where the time differential = T2 – T1.

    c. Determine the phase angle, where $\angle \theta = \dfrac{\text{time difference} \times 360°}{\text{time of one cycle}}$.

NOTE: If cursors #1 and #2 are interchanged, the results will be indicated by a minus (–) sign. It is also possible to measure the time of a complete cycle and the phase angle between the input and output waves by measuring between the 0.0 crossover points.

3. Insert the measured values, as indicated, into Table 15-6. Then, sketch the amplitude and phase shift of the V_{in} and V_o waveshapes on the graph of Figure 15-10(b) and label.

TABLE 15-6	f	X_C	Z	I	$V_R \angle \theta$	$V_C \angle \theta$
CALCULATE						
MEASURE						

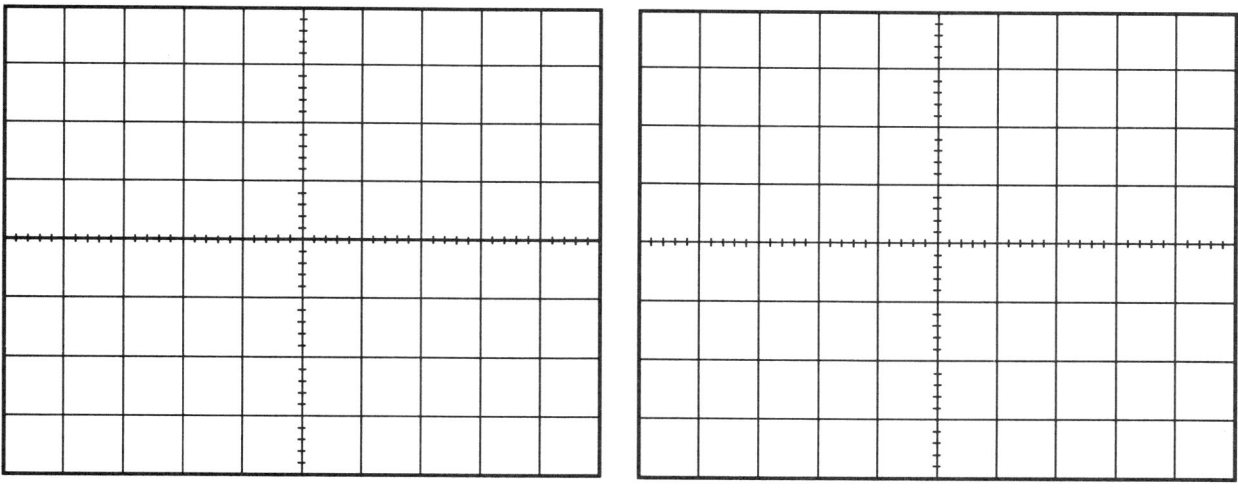

(a) V_{in} and V_C
1 V/div - 0.1 ms/div

(b) V_{in} and V_R
1 V/div - 0.1 ms/div

FIGURE 15-10

Questions and Problems: Basic Circuit Analysis for Electronics: 227-228.

CHAPTER 16
INDUCTORS IN AC SINE WAVE CIRCUITS

INTRODUCTION

Inductors are the third most widely used passive component in electronics circuits. Inductors, along with resistors and capacitors, are used in filter circuits that pass or reject ac signal voltages at selected frequencies. The formula for determining the inductive reactance of inductors is $X_L = 2\pi fL$, where f is the frequency in hertz and L is the inductance in henries. So, in the two inductor series circuit of Figure 16-1(a), X_{L1} and X_{L2} are solved, X_{LT} is solved, and then the Ohms law is used to solve I and the voltage drops across the L1 and L2 inductors. Then, in the series resistor and inductor circuit of Figure 16-1(b), X_L is solved, the impedance Z of the circuit is solved, and Ohm's law is used to solve the voltage drops across inductor L and resistor R.

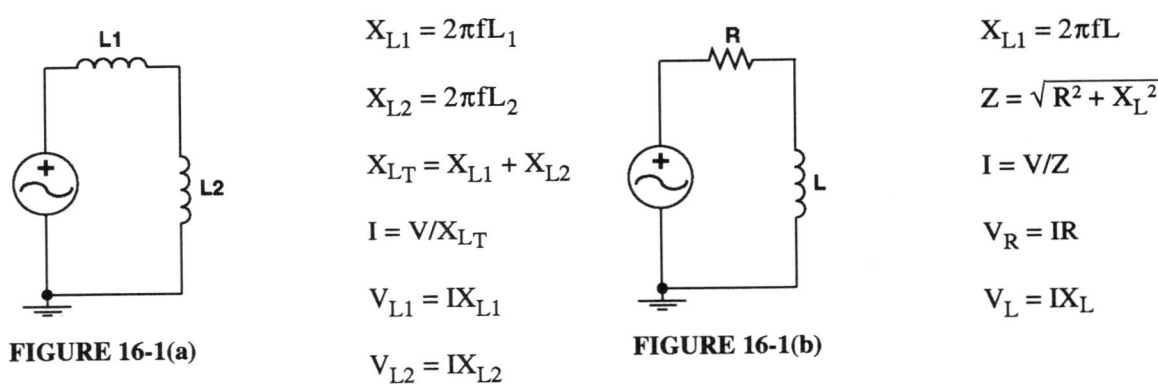

FIGURE 16-1(a)

$X_{L1} = 2\pi fL_1$

$X_{L2} = 2\pi fL_2$

$X_{LT} = X_{L1} + X_{L2}$

$I = V/X_{LT}$

$V_{L1} = IX_{L1}$

$V_{L2} = IX_{L2}$

FIGURE 16-1(b)

$X_{L1} = 2\pi fL$

$Z = \sqrt{R^2 + X_L^2}$

$I = V/Z$

$V_R = IR$

$V_L = IX_L$

In the resistor and two series inductor circuit of Figure 16-2(a), L_T and X_{LT} are solved, and the impedance Z of the circuit is solved. Then, Ohm's law is used to solve for I and the voltage drops across R and the series L1 and L2 inductors. In the resistor in series with the parallel inductors circuit of Figure 16-2(b), L_T and X_{LT} are solved. Then, the impedance Z of the circuit is solved and Ohms law is used to solve for I and the voltage drops across the R resistor and across the parallel L1 and L2 inductors.

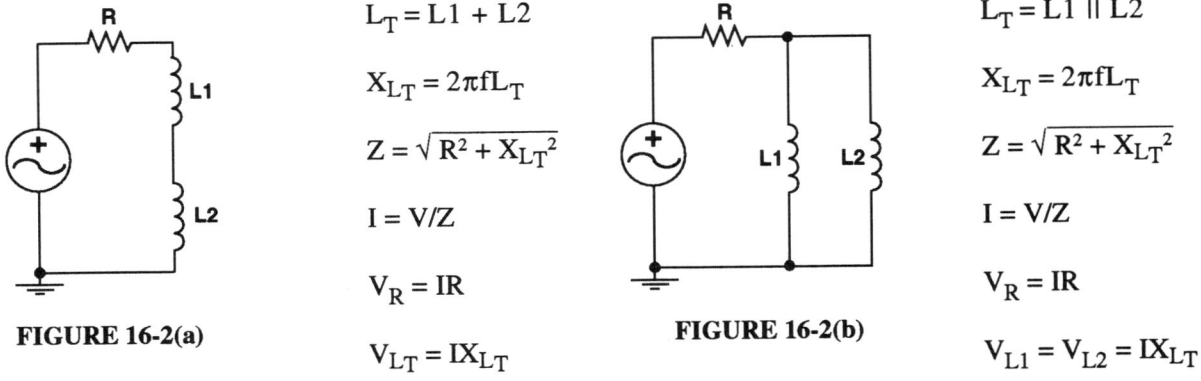

FIGURE 16-2(a)

$L_T = L1 + L2$

$X_{LT} = 2\pi fL_T$

$Z = \sqrt{R^2 + X_{LT}^2}$

$I = V/Z$

$V_R = IR$

$V_{LT} = IX_{LT}$

FIGURE 16-2(b)

$L_T = L1 \parallel L2$

$X_{LT} = 2\pi fL_T$

$Z = \sqrt{R^2 + X_{LT}^2}$

$I = V/Z$

$V_R = IR$

$V_{L1} = V_{L2} = IX_{LT}$

LABORATORY EXERCISE

READING ASSIGNMENT: Basic Circuit Analysis for Electronics: 229-236.
EXERCISE OBJECTIVES
To become familiar with :

- ac voltage divider circuits using series inductors.
- ac voltage divider circuits using series and parallel R and L components.
- Phase shift measurements.

PROCEDURE

SECTION I: Voltage Divider Circuits
PART A: Two Inductors in Series
Pre-Lab calculations
1. Analyze the two inductor series circuit of Figure 16-3(a), where L1 = 10 mH and L2 = 15 mH.

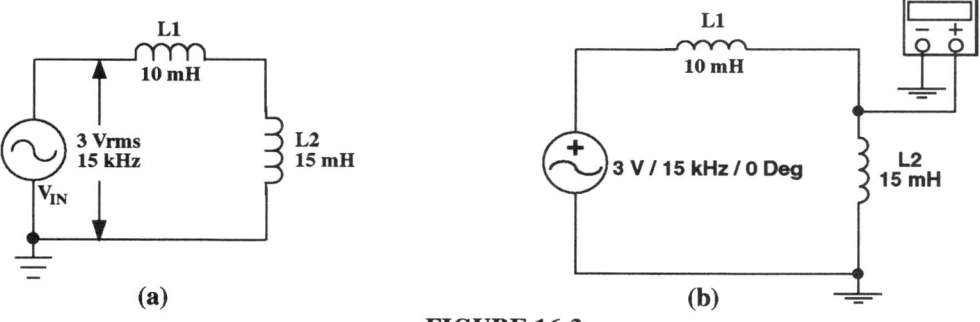

FIGURE 16-3

2. Calculate the output voltage drop across inductors L1 and L2, where V_{in} of the generator is set at 3 V rms and the frequency is 15 kHz.

   a. Solve the inductive reactance of the L1 = 10 mH and L2 = 15 mH inductors where:

   $X_{L1} = 2\pi f L_1$ and $X_{L2} = 2\pi f L_2$.

   b. Calculate the circuit current I, where $I = V_{in}/(X_{L1} + X_{L2})$.

   c. Calculate the ac voltage drop across the L1 inductor where: $V_{L1} = I(X_{L1})$.

   d. Calculate the ac voltage drop across the L2 inductor where: $V_{L2} = I(X_{L2})$.

3. Insert the calculated values, as indicated, into Table 16-1.

Voltmeter Circuit Measurements

Open and select the circuit diagram of Figure 16-3(b) from the File menu , or connect the circuit. Set the multimeter to ac (~) and volts (V). The generator is set 3 V rms at 15 kHz.

1. Measure V_{L2}, the ac voltage drop across the L2 inductor.

2. Then, find V_{L1}, the ac voltage drop across the L1 inductor indirectly where: $V_{L1} = V_{in} - V_{L2}$.

3. Insert the measured values, as indicated, into Table 16-1.

TABLE 16-1	X_{L1}	X_{L2}	I	V_{L1}	V_{L2}
CALC.					
MEAS.	/////	/////	/////		

PART B: Resistor and Inductor Series Circuit
Pre-Laboratory Calculations
1. Analyze the RL series circuit of Figure 16-4(a), where R = 1.2 kΩ and L = 10 mH.

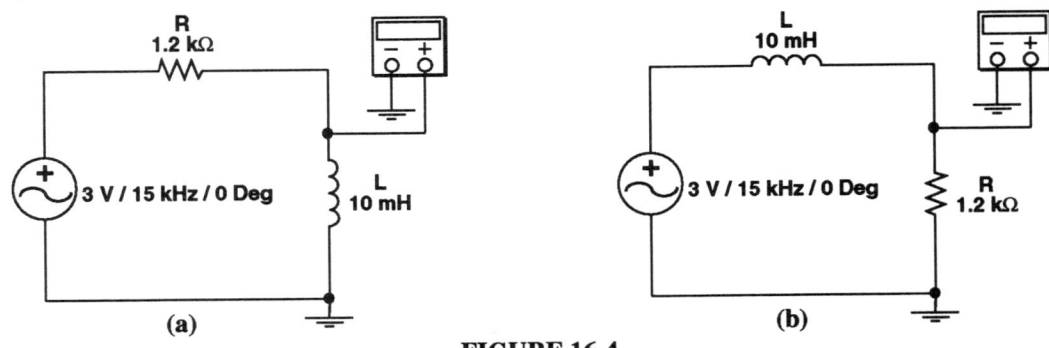

FIGURE 16-4

2. Calculate the output voltage drop across inductor L, where V_{in} of the generator is set at 3 rms and 15 kHz.

   a. Solve the inductive reactance of the L = 10 mH inductor where: $X_L = 2\pi fL$.

   b. Calculate the impedance Z, of the R and X_L series circuit where: $Z = \sqrt{R^2 + X_L^2}$.

   c. Calculate the circuit current I, where $I = V_{in}/Z$.

   d. Calculate the ac voltage drop across the R resistor where: $V_R = IR$.

   e. Calculate the ac voltage drop across the L inductor where: $V_L = I(X_L)$.

3. Insert the calculated values, as indicated, into Table 16-2.

Voltmeter Circuit Measurements
Open and select the circuit of Figure 16-4(a) from the File menu, or connect the circuit. Set the multimeter to ac (~) and volts (V). Set the V_{in} of the generator at 3 V rms and the frequency at 15 kHz.
1. Measure V_L, the ac voltage drop across the L inductor.

2. Interchange the resistor and inductor, as shown in Figure 16-4(b), and maintain the V_{in} of the generator at 3 V rms and the frequency at 15 kHz. Measure the ac voltage drop across the resistor R.

3. Use the measured V_R and V_L to provide X_L and Z.

   a. Use the measured V_R and the R value to determine the circuit current where: $I = V_R/R$.

   b. Use the measured V_L and the circuit current I to determine X_L where: $X_L = V_{XL}/I$.

   c. Use V_{in} and the circuit current I to determine Z where: $Z = V_{in}/I$.

4. Insert the measured values, as indicated, into Table 16-2.

TABLE 16-2	X_L	Z	I	V_R	V_L
CALC.					
MEAS.					

SECTION II: Series and Parallel Inductor Circuits
PART A: Resistor in Series with Parallel Inductors Circuit
Pre-Laboratory Calculations
1. Analyze the resistor R in series with the parallel inductors L1 and L2 in the circuit of Figure 16-5(a), where R = 1.2 kΩ, L1 = 10 mH, and L2 = 15 mH.

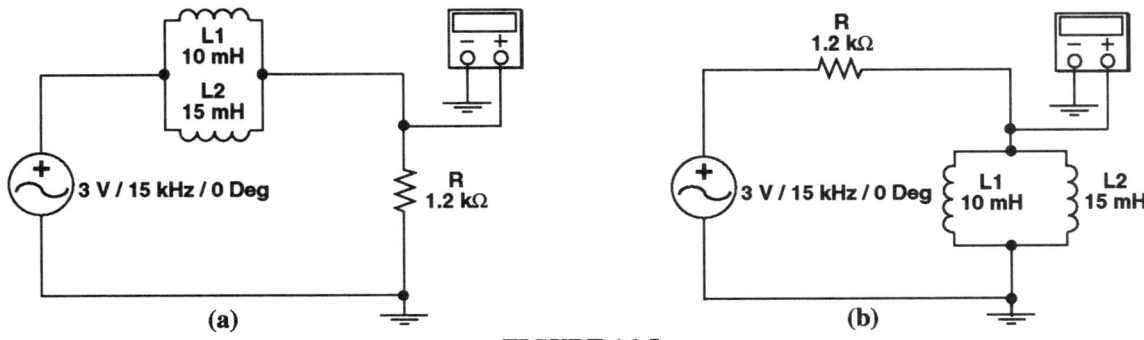

FIGURE 16-5

2. Calculate the output voltage drop across the parallel L1 and L2 inductors, where V_{in} of the generator is set at 3 V rms and 15 kHz.
   a. Solve L_T from $L_T = L1 \parallel L2$, where: L1 = 10 mH and L2 = 15 mH.
   b. Solve the inductive reactance of the L_T where: $X_{LT} = 2\pi f L_T$.
   c. Calculate the impedance of the R and X_{LT} series circuit where: $Z = \sqrt{R^2 + X_{LT}^2}$.
   d. Calculate the circuit current I, where $I = V_{in}/Z$.
   e. Calculate the ac voltage drop across the L1 ∥ L2 inductors where: $V_{LT} = I(X_{LT})$.
   f. Calculate the ac voltage drop across R where: $V_R = IR$.
3. Insert the calculated values, as indicated, into Table 16-3.

Voltmeter Circuit Measurements
Open and select the circuit diagram of Figure 16-5(a) from the File menu, or connect the circuit. The multimeter is set to ac (~) and volts (V). Set the generator V_{in} to 3 V rms and the frequency to 15 kHz.

1. Measure V_{LT}, the ac voltage drop across the parallel L1 and L2 inductors.

2. Interchange the resistor and parallel inductors, as shown in Figure 16-5(b), and maintain V_{in} of the generator at 3 V rms and the frequency at 15 kHz. Measure the ac voltage drops across the R resistor.
   a. Use the measured V_R and the R value to determine the circuit current where: $I = V_R/R$.
   b. Use the measured V_{LT} and the circuit current to determine X_{LT} where: $X_{LT} = V_{LT}/I$.
   c. Use X_{LT} to determine L_T where: $L_T = X_{LT}/(2\pi f)$.
   d. Use V_{IN} and the circuit current I to determine Z where: $Z = V_{in}/I$.

3. Insert the measured values, as indicated, into Table 16-3.

TABLE 16-3	L_T	X_{LT}	Z	I	V_{LT}	V_R
CALCULATED						
MEASURED						

PART B: Resistor R and Series L1 and L2 Inductor Circuit
Pre-Laboratory Calculations
1. Analyze the circuit of Figure 16-6(a), where R = 1.2 kΩ, L1 = 10 mH, and L2 = 15 mH.
2. Calculate the V_o drop across series L1 and L2, where V_{in} of the generator is set at 3 V rms and 15 kHz.

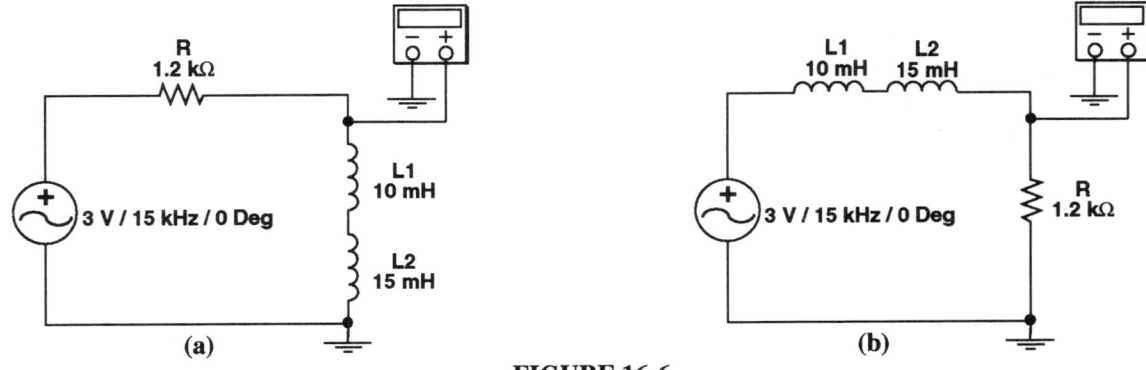

FIGURE 16-6

a. Solve L_T where: $L_T = L1 + L2$ and $L1 = 10$ mH and $L2 = 15$ mH
b. Solve the inductive reactance of L_T where: $X_{LT} = 2\pi f L_T$.
c. Calculate the impedance of the R and X_{LT} series circuit where: $Z = \sqrt{R^2 + X_{LT}^2}$.
d. Calculate the circuit current I where: $I = V_{in}/Z$
e. Calculate the ac voltage drop across the series L1 and L2 inductors where: $V_{LT} = I(X_{LT})$.
f. Calculate the ac voltage drop across R where: $V_R = IR$.

3. Insert the calculated values, as indicated, into Table 16-4.

Voltmeter Circuit Measurements

Open and select the circuit diagram of Figure 16-6(a) from the File menu, or connect the circuit. The multimeter is set to ac (~) and volts (V). Set the V_{in} of the generator at 3 V rms and the frequency at 15 kHz.

1. Measure V_{LT}, the ac voltage drop across the parallel L1 and L2 inductors.

2. Interchange the resistor and series inductors, as shown in Figure 16-6(b) Maintain V_{in} of the generator at 3 V rms and the frequency at 15 kHz. Measure the ac voltage drops across the resistor R.

3. Use the measured V_R and V_{LT} to prove X_{LT} and L_T.
   a. Use the measured V_R and the R value to determine the circuit current where: $I = V_R/R$.
   b. Use the measured V_{LT} and the circuit current to determine X_{LT} where: $X_{LT} = V_{LT}/I$.
   c. Use X_{LT} to determine series L_1 and L_2 inductors where: $L_T = X_{LT}/(2\pi f)$.
   d. Use V_{IN} and the circuit current I to determine Z where: $Z = V_{in}/I$.

4. Insert the measured values, as indicated, into Table 16-4.

TABLE 16-4	L_T	X_{LT}	Z	I	V_{LT}	V_R
CALCULATED						
MEASURED						

SECTION III: Oscilloscope Phase Shift Measurements
PART A. Pre-Lab calculations

1. Analyze the RC series circuit of Figure 16-7(a), where $X_L = R$, $R = 1.2$ kΩ, and $L = 10$ mH.

2. Calculate the output voltage drop across inductor L, where V_{in} of the generator is set at 6 Vp-p at the operating frequency where $X_L = R$.
   a. Calculate the operating frequency where: $f = R/(2\pi L)$.
   b. Calculate the inductive reactance of the $L = 10$ mH inductor where: $X_L = 2\pi f L$.
   c. Calculate the impedance Z, of the R and X_L series circuit where: $Z = \sqrt{R^2 + X_L^2}$.
   d. Calculate the phase angle $\angle\theta$ where: $\angle\theta = \tan^{-1} X_L/R$.

Inductors in AC Sine Wave Circuits — 117

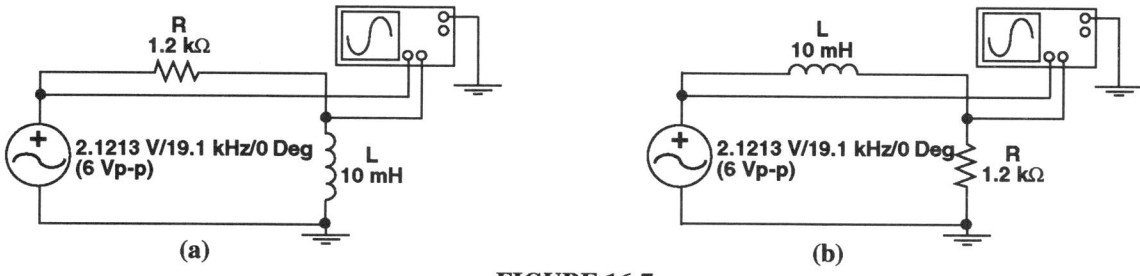

FIGURE 16-7

  e. Calculate the circuit current I where: $I = V/Z\angle\theta$.
  f. Calculate the ac voltage drop across the R resistor where: $V_R = I\angle\theta(R)$.
  g. Calculate V_o, the ac voltage drop across the L inductor where: $V_o = I\angle\theta(X_L\angle\theta)$.

3. Insert the calculated values, as indicated, into Table 16-5.

PART B: Oscilloscope Measurements of Figure 16-7(a)

Open and select the circuit of Figure 16-7(a) from the File menu, or connect the circuit. Set the V_{in} of the generator to 6 Vp-p (2.1213 V rms) at the calculated frequency, where $X_L = R$. The dual trace scope is set to the ac mode and to 1V/div and 20 µs/div.

1. Amplitude Measurements: Measure the amplitude of the input reference voltage V_{in} and the amplitude of the output voltage V_o across the L inductor.
    a. Measure the amplitude of the input voltage V_{in}. Click to enlarge the scope and click on the power switch to activate the circuit. Select Expand and use cursor #2 to monitor the positive-going wave and cursor #1 to monitor the negative-going wave. Read the Vp-p swing directly where: $V_{in} = VA1 - VA2$.
    b. Measure the amplitude of the smaller in amplitude output voltage V_o following the procedure of Step a. Use cursor #2 to monitor the positive-going wave of the phase leading wave (≈ 0.707 of V_{in}) and cursor #1 to monitor the negative-going wave. Read the peak-to-peak voltage directly.

2. Phase Measurements: Measure the time of one cycle and measure the time difference between V_{in} and V_o.
    a. Measure the time of one cycle by using cursor #1 and cursor #2 to monitor between the positive-going peaks of a complete cycle and read the time directly from T2 - T1.
    b. Measure the time between the input and output waves (time differential). Set cursor #1 at the positive-going peak of the phase-leading wave and cursor #2 at the positive-going peak voltage of V_{in}. Read the time differential between the two waves and read the time directly where: t diff. = T2 - T1.
    c. Determine the phase angle, where $\angle\theta = \dfrac{\text{t difference} \times 360°}{\text{time of one cycle}}$.

NOTE: If cursors #1 and #2 are interchanged, the results will be indicated by a minus (−) sign. It is also possible to measure the time of a complete cycle and the phase angle between the input and output waves by measuring between the 0.0 crossover points.

3. Insert the measured values, as indicated, into Table 16-5. Then, sketch the amplitude and phase shift of the V_{in} and V_o waveshapes on the graph of Figure 16-8(a)

PART C: Scope Circuit Measurements of Figure 16-7(b)

Interchange the R and L components, as shown in Figure 16-7(b), and monitor the voltage and phase angle across the resistor output. Maintain the generator and scope settings.

1. Amplitude Measurements: Measure the amplitude of the input reference voltage V_{in} and the amplitude of the output voltage V_o across the L inductor.
    a. Measure the amplitude of the input reference voltage V_{in}. Click to enlarge the scope and click on the power switch to activate the circuit. Select Expand and use cursor #2 to monitor the positive-going vave and cursor #1 to monitor the negative-going wave. Read the Vp-p swing directly from $V_{in} = VA1 - VA2$.
    b. Measure the amplitude of the smaller in amplitude output voltage V_o following the above procedure.

Use cursor #2 to monitor the positive-going wave of the phase leading wave (≈ 0.707 of V_{in}) and use cursor #1 to monitor the negative-going wave. Read the peak-to-peak voltage directly.

2. Phase Measurements: Measure the time of one cycle and measure the time difference between V_{in} and V_o.
   a. Measure the time of one cycle by using cursor #1 and cursor #2 to monitor between the positive-going peaks of a complete cycle and read the time directly where: t diff. = T2 – T1.
   b. Measure the time between the input and output waves (time differential). Set cursor #1 at the positive-going peak voltage of V_{in} and cursor #2 at the positive-going peak of the phase leading wave. Read the time differential between the two waves and read the time directly where t diff. = T2 – T1.
   c. Determine the phase angle, where $\angle\theta = \dfrac{\text{t difference} \times 360°}{\text{time of one cycle}}$.

NOTE: If cursors #1 and #2 are interchanged, the results will be indicated by a minus (–) sign. It is also possible to measure the time of a complete cycle and the phase angle between the input and output waves by measuring between the 0.0 crossover points.

3. Insert the measured values, as indicated, into Table 16-5. Then, sketch the amplitude and phase shift of the V_{in} and V_o waveshapes on the graph of Figure 16-8(b)

TABLE 16-5	f	X_L	Z	I	$V_R \angle \theta$	$V_L \angle \theta$
CALCULATED						
MEASURED	/////	/////	/////	/////		

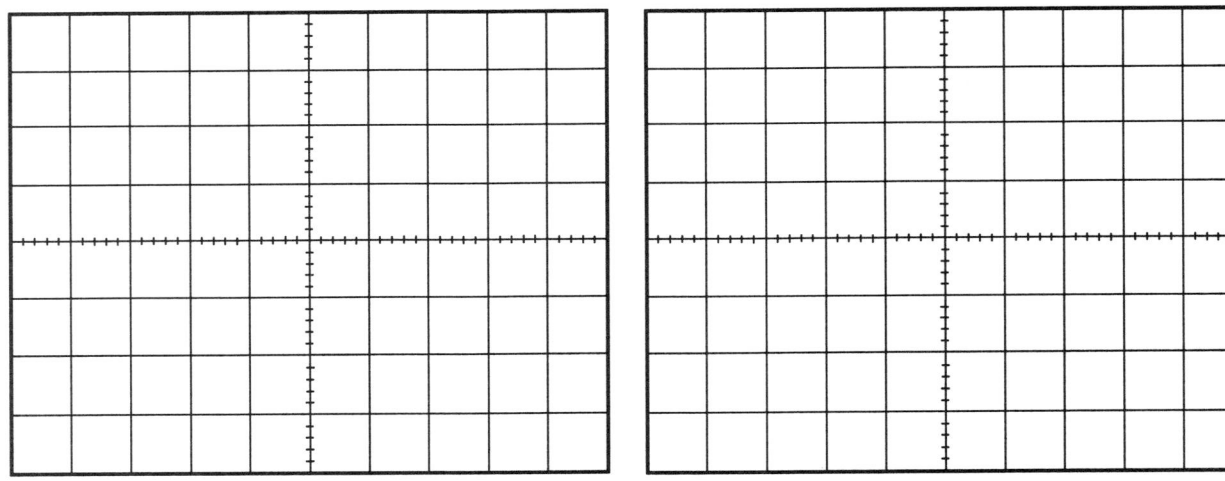

(a) V_{in} and V_L (1 V/ div - 20 µs/div) (b) V_{in} and V_R (1 V/div - 20 µs/div)

FIGURE 16-8

Questions and Problems: Basic Circuit Analysis for Electronics: 243.

CHAPTER 17
TRANSFORMERS

INTRODUCTION

Transformers step down or step up the input voltage and they provide isolation. They are essentially two inductors in close proximity. When an input signal is applied to the primary winding, the current flow into the winding causes a flux field to surround both the primary and secondary windings. This causes voltage to be developed across the secondary and current to flow into the load resistor.

Ideal transformers are assumed in the EWB program so no power loss occurs between windings, and the coefficient of coupling is unity (k =1). Therefore the turns ratio of the primary and secondary windings (N) can be used to solve the ratio of output to input voltage, or $N = V_o/V_{in} = N_P/N_S$. In practice, N can be found by using the ratio of the input to output voltage, or by using the measured primary inductance (L_{pri}) and secondary inductance (L_{Sec}) to solve N from: $N = \sqrt{L_{Pri}/L_{Sec}}$. Transformers also can provide an output voltage that is in or out of phase with the input sinewave voltage. Also, since the power developed at the load of an ideal transformer equals the power delivered to the input, the effect of the reflected resistance to the input can be solved from either $r_L = V_{in}/I_{in}$, or from $r_L = N^2 \times R_L$. The formulas are shown in conjunction with the circuit of Figure 17-1.

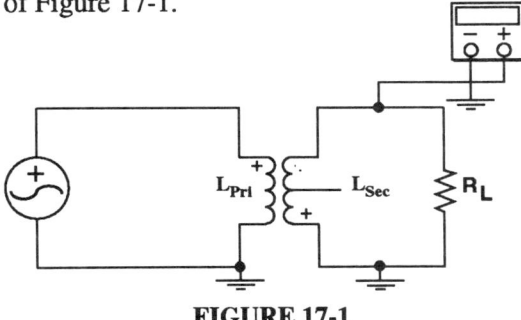

FIGURE 17-1

$$N = N_p/N_s = \sqrt{L_{Pri}/L_{Sec}}$$

$$V_o = \frac{V_{in}N_s}{N_p} = \frac{V_{in}}{N}$$

$$P_{RL} = \frac{V_o^2}{R_L}$$

$$I_{RL} = \frac{P_{RL}}{V_o}$$

$$I_{in} = \frac{P_{in}}{V_{in}}$$

$$r_L = \frac{V_{in}}{I_{in}}$$

$$r_L = N^2 R_L$$

AUTOTRANSFORMERS

Autotransformers effectively step-down or step-up the line voltage using a single winding, as shown in Figure 17-2. In theory, in an ideal 2:1 step-down transformer, the primary windings (N_P) will have twice as many turns as the secondary windings (N_S). So the input line voltage connected across the full winding (N1 + N2) and taken off the center tap (N2) will step-down the input line voltage 2 : 1, as shown in Figure 17-2. N can also be solved from: $N = \sqrt{L_{Sec}/L_{Sec-2}}$.

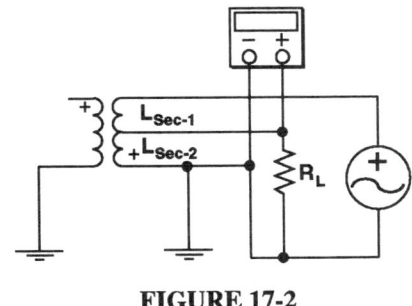

FIGURE 17-2

$N = (N1 + N2)/N2$, or

$N = \sqrt{(L_{Sec}/L_{Sec-2}}$

$V_o = V_{in}/N$

where: $N1 = L_{Sec-1}$
$N2 = L_{Sec-2}$

LABORATORY EXERCISE

READING ASSIGNMENT: Basic Circuit Analysis for Electronics: 244-250.

EXERCISE OBJECTIVES
To become familiar with:

- Inductor measurements of transformers.
- Step-down transformers.
- Step-up transformers.
- Autotransformers.

PROCEDURE

SECTION I: Transformer Inductance Measurements

PART A: Measuring the Primary Inductance

Use the circuit of Figure 17-3(b) to indirectly measure the primary inductance (L_{Pri}) which is measured across terminals 1 and 2. Select Figure 17-3(b) from the File menu and open, or connect the circuit. In the connection the ac constant current source is set to 20 mA and 100 Hz and the multimeter is set to ac (~) and volts (V).

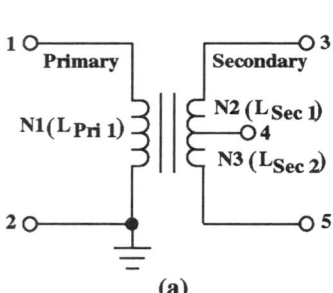
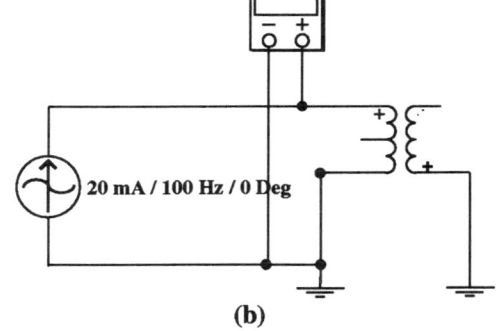

FIGURE 17-3

1. Activate the circuit and measure V_{Pri}, the ac voltage across the primary of the transformer.

2. Solve the X_L inductance where: $X_{L(Pri)} = V_{Pri}/I$ and where I = 20 mA.

3. Once $X_{L(Pri)}$ is known, find the inductance of the primary where: $L_{Pri} = X_{L(Pri)}/2\pi f$.

4. Insert the calculated measured values, as indicated, into Table 17-1.

PART B: Measuring the Secondary Inductance

The circuit of Figure 17-4 is used to measure the secondary inductance (L_{Sec}) across the secondary winding terminals of 3 and 5. Open and select the circuit from the File menu, or connect the circuit. In the connection the ac constant current source is set to 20 mA and 100 Hz and the multimeter is set to ac (~) and volts (V).

1. Activate the circuit and measure V_{Sec}, the ac voltage across the secondary of the transformer.

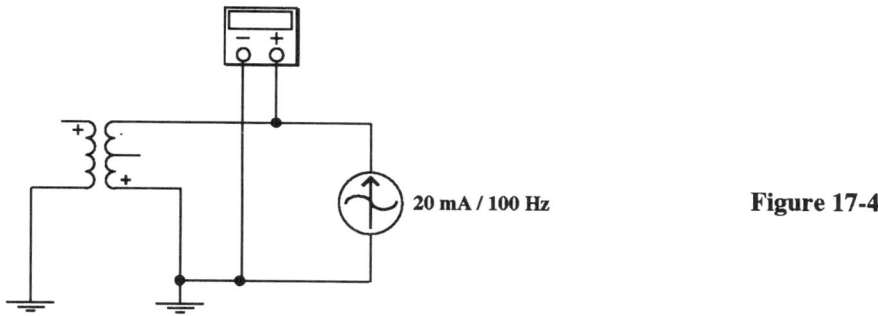

Figure 17-4

2. Solve $X_{L(Sec)}$ indirectly where: $X_{L(Sec)} = V_{Sec}/I$ and where I = 20 mA.

3. Once $X_{L(Sec)}$ is known, the inductance of the primary can be found from $L_{(Sec)} = X_{L(Sec)}/2\pi f$.

4. Insert the calculated and measured values, as indicated, into Table 17-1.

PART C: Measuring the L_{Sec-1} and L_{Sec-2} Secondary Inductance

The circuits of Figure 17-5 are used to measure each of the secondary windings of the transformer. L_{Sec-1} is measured across terminals 3 and 4 and L_{Sec-2} is measured across terminals 4 and 5. In turn, select the Figures from the File menu and open, or connect them. The ac constant current source is set to 20 mA and 100 Hz and the multimeter is set to ac (~) and volts (V).

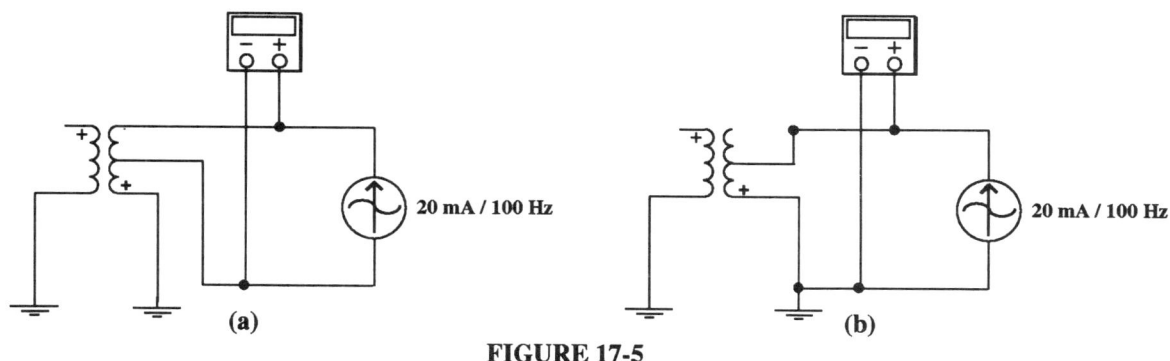

FIGURE 17-5

1. Measure L_{Sec-1} directly, as shown in Figure 17-5(a), to find $V_{L(Sec-1)}$.

2. Use the $V_{L(Sec-1)}$ to find the $X_{L(Sec-1)}$ inductance where: $X_{L(Sec-1)} = V_{Sec-1}/I$.

3. Repeat Steps 1 and 2 to find $V_{L(Sec-2)}$ and $X_{L(Sec-2)}$. Refer to Figure 17-5(b)

4. Verify the secondary inductance from: $L_{Sec} = L_{Sec-1} + L_{Sec-2} + 2 M_{Sec}$ where $M_{Sec} = \sqrt{L_{Sec-1}L_{Sec-2}}$.

5. Insert the calculated and measured values, as indicated, into Table 17-1.

TABLE 17-1	$L_{Primary}$		$L_{Secondary}$		L_{Sec-1}		L_{Sec-2}		Verification	
	$V_{(Pri)}$	L_{Pri}	$V_{(Sec)}$	L_{Sec}	$V_{(Sec-1)}$	L_{Sec-1}	$V_{(Sec-2)}$	L_{Sec-2}	L_{Sec}	M_{Sec}
CALC.										
MEAS.										

SECTION II: STEP-DOWN TRANSFORMER ACTION
PART A: PRIMARY TO FULL SECONDARY

Open and select the circuit of Figure 17-6 from the File menu, or connect the circuit. Set the input signal voltage to 3 V rms at 100 Hz. The input signal is applied to the transformer primary and the output signal is developed across the load resistor R_L.

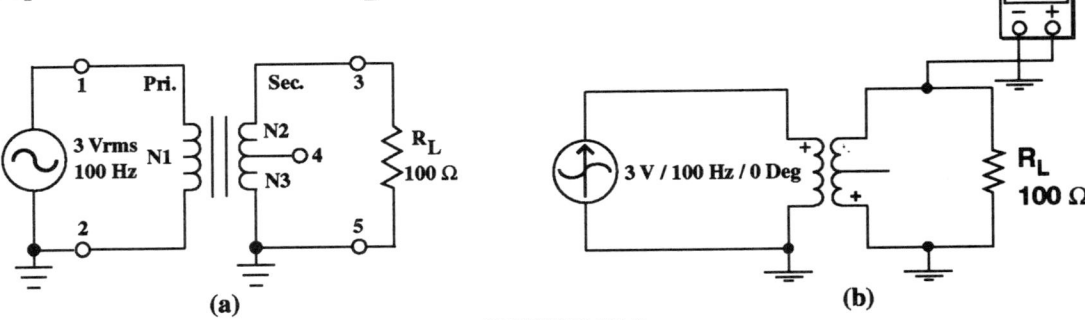

FIGURE 17-6

1. Calculate the output voltage across R_L.
   a. Use the primary and secondary winding inductances to calculate the turns ratio of the primary to the secondary windings where: $N = \sqrt{L_{Pri}/L_{Sec}}$.
   b. Calculate the voltage across the load resistor where: $V_o = V_{in}/N$.

2. Measure the voltage across R_L.
   a. Check to see that the input voltage is set to 3 V rms at 100 Hz.
   b. Measure the voltage across the secondary and load resistor R_L.

3. Insert the calculated and measured values, as indicated, into Table 17-2.

PART B: Primary to One-Half of the Secondary

Open and select the circuit of Figure 17-7 from the File menu, or connect the circuit. The input voltage is set to 3 V rms at 100 Hz. The signal is applied to the primary and developed across one-half of the secondary and load resistor R_L.

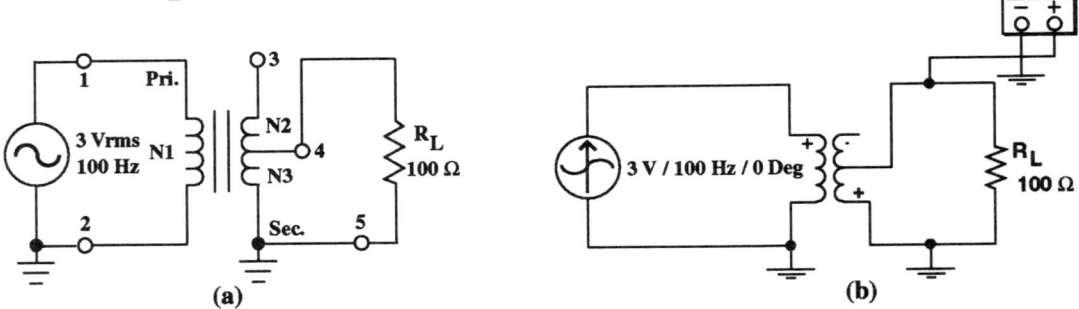

FIGURE 17-7

1. Calculate the voltage across R_L.
   a. Use the measured primary and secondary inductances of the transformer to calculate the turns ratio N of the primary to one-half the secondary winding where: $N = \sqrt{L_{Pri}/L_{Sec-2}}$.
   b. Calculate the voltage across the load resistor R_L where: $V_o = V_{in}/N$.

2. Measure the output voltage across R_L.
   a. Check to see that the input voltage is set to 3 V rms at 100 Hz.
   b. Measure the voltage across the secondary and load resistor R_L.

3. Insert the calculated and measured values, as indicated, into Table 17-2.

TABLE 17-2	Primary to Full Secondary			Primary to One-Half Secondary		
	V_{in}	N	V_{RL}	V_{in}	N	V_{RL}
CALCULATE						
MEASURE						

SECTION III: Step-Up Transformer Action

PART A: Full Secondary to Primary

Open and select the circuit of Figure 17-8 from the File menu, or connect the circuit. The input signal voltage is 300 mV rms at 100 Hz. The input signal is applied to the full secondary and the output signal is developed across the primary and load resistor R_L.

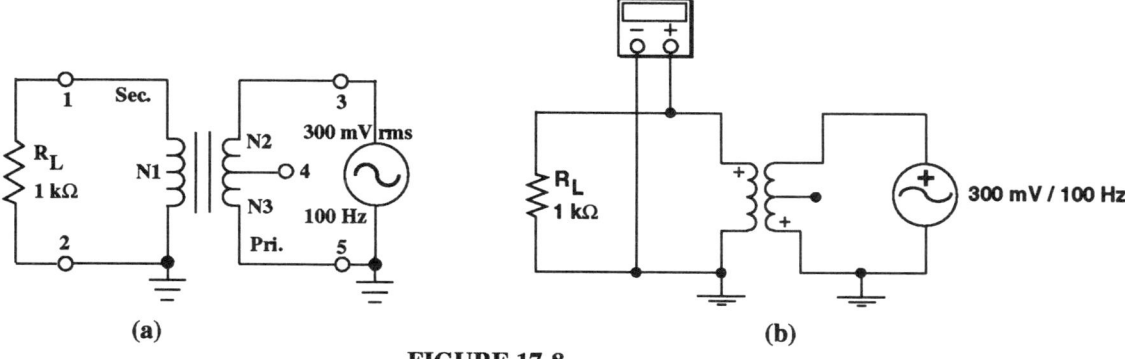

FIGURE 17-8

1. Calculate the output voltage across R_L.
   a. Use the measured primary and secondary winding inductances to calculate the turns ratio of the secondary to primary windings where: $N = \sqrt{L_{Sec}/L_{Pri}}$.
   b. Calculate the voltage across the load resistor where: $V_o = V_{in}/N$.

2. Measure the output voltage across R_L.
   a. Check to make sure the input voltage is set at 300 mV rms and 100 Hz.
   b. Measure the voltage across the primary and the load resistor R_L.

3. Insert the calculated and measured values, as indicated, into Table 17-3.

PART B: One-Half Secondary to Primary

Open and select the circuit of Figure 17-9 from the File menu, or connect the circuit. The input signal voltage is 300 mV rms at 100 Hz. The input signal is applied to one-half the secondary and the output signal is developed across the primary and load resistor R_L.

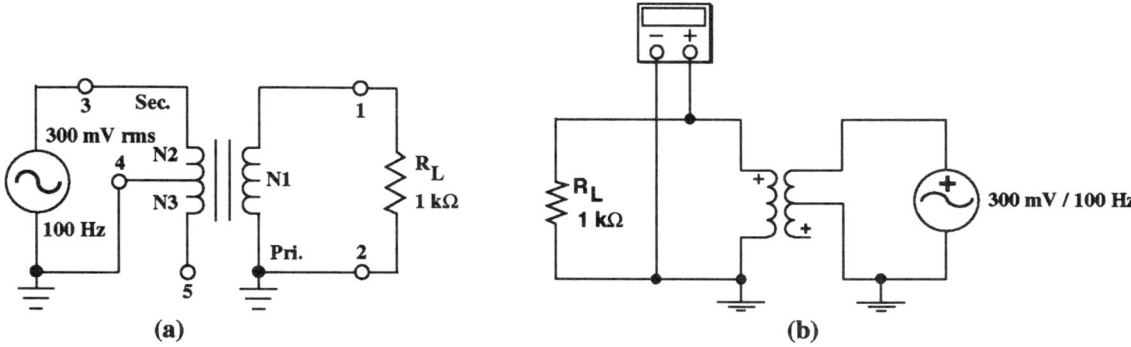

FIGURE 17-9

1. Calculate the output voltage across R_L.
   a. Use the measured one-half secondary and primary winding <u>inductances</u> to calculate the turns ratio N of the one-half secondary to primary windings where: $N = \sqrt{L_{Sec-2}/L_{Pri}}$.
   b. Calculate the voltage across the load resistor where: $V_o = V_{in}/N$.

2. Measure the output voltage across R_L.
   a. Check to make sure the input voltage is set at 300 mV rms and 100 Hz.
   b. Measure the voltage across the primary and the load resistor R_L.

3. Insert the calculated and measured values, as indicated, into Table 17-3.

TABLE 17-3	Full Secondary to Primary			One-Half Secondary to Primary		
	V_{in}	N	V_{RL}	V_{in}	N	V_{RL}
CALCULATE						
MEASURE						

SECTION IV: Autotransformer Action
PART A: Step-Down Transformer

Open and select the circuit of Figure 17-10 from the File menu, or connect the circuit. Set the input signal voltage to 3 V rms at 100 Hz. The input signal is applied to the full secondary and the output signal is developed across one-half the secondary and the load resistor R_L. The primary windings are not used and are not connected (floating).

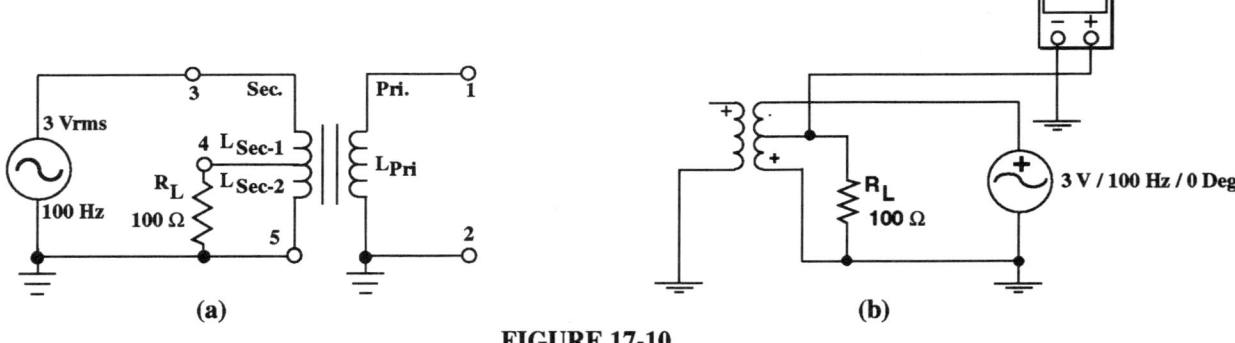

FIGURE 17-10

1. Calculate the output voltage across R_L.
   a. Use the measured one-half secondary and full secondary winding inductances to calculate the turns ratio of the primary to the secondary windings where: $N = \sqrt{L_{Sec}/L_{Sec-1}}$.
   b. Calculate the voltage across the load resistor where: $V_o = V_{in}/N$.

2. Measure the voltage across R_L.
   a. Check to see that the input voltage is set to 3 V rms at 100 Hz.
   b. Measure the voltage across the secondary (V_{Sec-2}) and load resistor R_L.

3. Insert the calculated and measured values, as indicated, into Table 17-4.

PART B: Step-Up Transformer

Open and select the circuit of Figure 17-11 from the File menu, or connect the circuit. The input voltage is set to 3 V rms at 100 Hz. The signal is applied to one-half the secondary and developed across the full secondary and load resistor R_L. Again, the primary windings are not used and are not connected (floating).

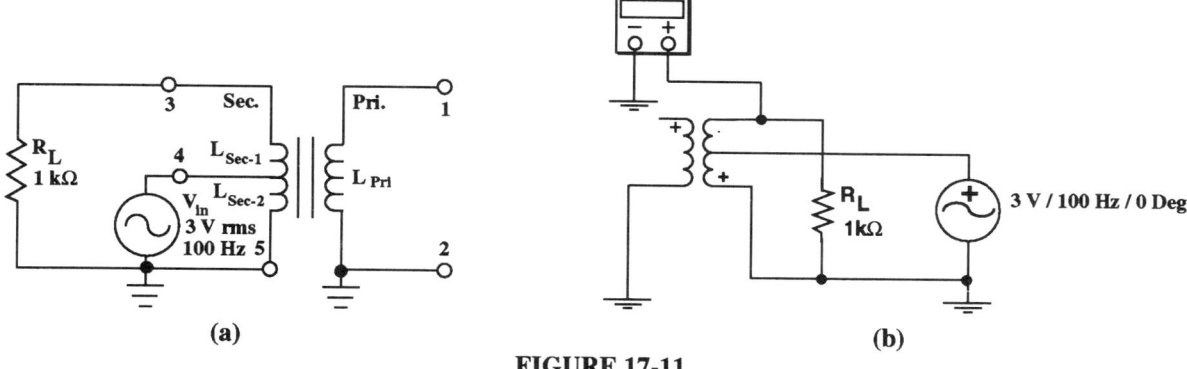

FIGURE 17-11

1. Calculate the output voltage across R_L.
   a. Use the measured one-half secondary and the full secondary winding inductances to calculate the turns ratio N of the one-half secondary and the full secondary windings where: $N = \sqrt{L_{Sec-2}/L_{Sec}}$.
   b. Calculate the voltage across the load resistor where: $V_o = V_{in}/N$.

2. Measure the voltage across R_L.
   a. Check to see that the input voltage is set to 3 V rms at 100 Hz.
   b. Measure the voltage across the secondary and load resistor R_L.

3. Insert the calculated and measured values, as indicated, into Table 17-4.

TABLE 17-4	Step-Down Autotransformer			Step-Up Autotransformer		
	V_{in}	N	V_{RL}	V_{in}	N	V_{RL}
CALCULATE						
MEASURE	/////	/////		/////	/////	/////

SECTION V: Phase Measurements
PART A

In the connection of Figure 17-12(a) a dual trace oscilloscope is used to measure the phase condition between the primary and secondary windings. Winding 2 of the primary and winding 5 of the secondary are the reference grounds for the connection. Use an input signal of 6 Vp-p at 125 Hz and set the scope at 1 V/div and 1 ms/div.

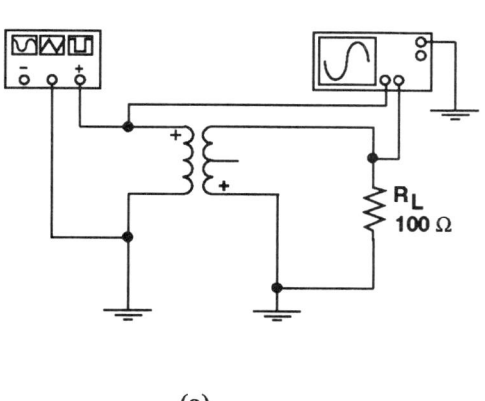
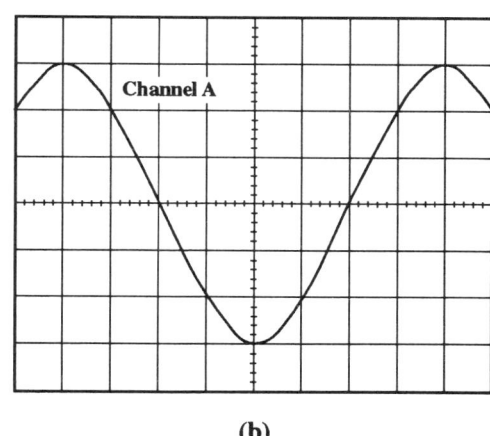

FIGURE 17-12

1. Use Channel A (or channel 1) to monitor the primary terminals 1 and 2. Make sure the ground side of the scope lead is connected to terminal 2 and the high side (probe) connected to terminal 1.

2. Use Channel B (or channel 2) to monitor the secondary terminals 3 and 5 and the load. Make sure the ground side of the scope lead is connected to terminal 5 and the high side (probe) connected to terminal 3.

3. Monitor the input and output signals and sketch the waveshapes and phase difference, if any, into the graph of Figure 17-13(b).

PART B

In the connection of Figure 17-13(a), a dual trace scope is once more used to measure the phase condition between the primary and secondary. However, in this connection terminal 2 of the primary and terminal 3 of the secondary are the reference grounds. Use an input signal of 6 Vp-p at 125 Hz and set the scope at 1 V/div and 1 ms/div.

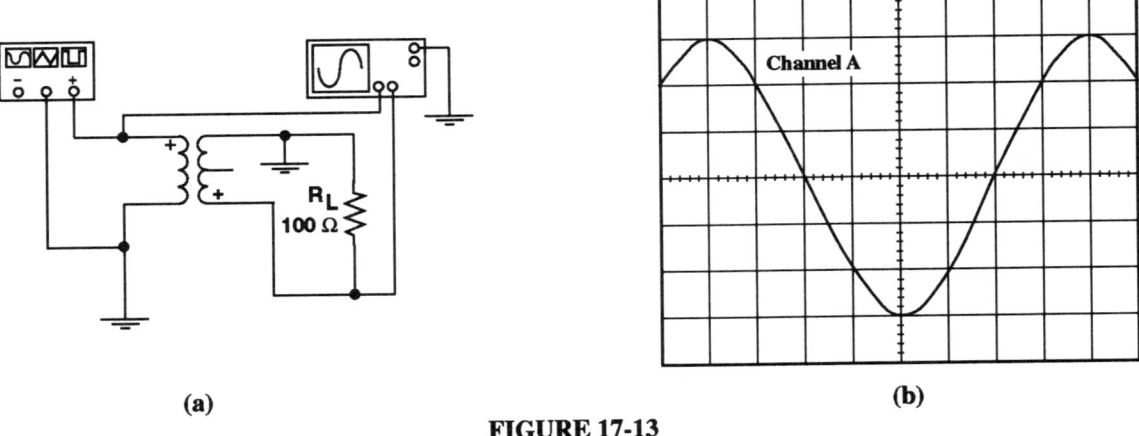

(a)   (b)

FIGURE 17-13

1. Use Channel A (or channel 1) to monitor the primary terminals 1 and 2. Make sure the ground side of the scope lead is connected to terminal 2 and the high side (probe) connected to terminal 1.

2. Use Channel B (or channel 2) to monitor the secondary terminals 3 and 5 and the load. Make sure the ground side of the scope lead is connected to terminal 3 and the high side (probe) connected to terminal 5.

3. Monitor the input and output signals and sketch the waveshapes and phase difference, if any, into the graph of Figure 17-13 (b).

Questions and Problems: Basic Circuit Analysis for Electronics: 257-258

CHAPTER 18
LOW PASS FILTERS

INTRODUCTION

Low pass filters, in their most basic form, can be constructed using either a resistor and a capacitor, or a resistor and an inductor. In the basic two component RC low pass filter circuit of Figure 18-1, the resistance remains fixed but the capacitive reactance decreases as the frequency is increased. As the frequency is increased the output voltage decreases. The standard formulas and transfer function formulas for solving the output voltage at any given frequency are shown in conjunction with the figure. The advantage of transfer functions is that the formula is easily applied. For the RC low pass filter the formula (which is the same for RL circuit) equals $A = 1/1 + j(f_o/f_c)$. So, begin by solving the corner frequency. Then plug both the corner frequency and each of the operating frequencies into the formula to solve A, the phase angle, and the gain (A) in dB.

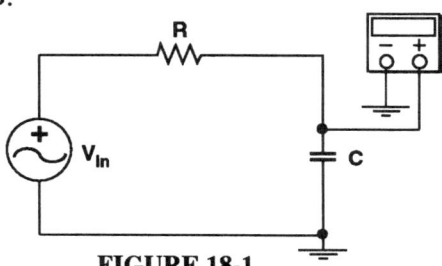

$$X_C = \frac{1}{2\pi fC}$$

$$Z = \sqrt{R^2 + jX_C^2}$$

$$I = V_{in}/Z$$

$$V_C = IX_C$$

$$A(dB) = 20 \log V_{XC}/V_{in}$$

$$f_C = \frac{1}{2\pi RC}$$

$$A = \frac{1}{1 + j(f_o/f_c)}$$

$$\angle\theta = \tan^{-1}(f_o/f_c)$$

$$A(dB) = 20 \log A$$

FIGURE 18-1   **Standard Formulas**   **Transfer Function Formulas**

In the basic two component RL circuit of Figure 18-2, the inductive resistance of the inductor increases as the frequency is increased and, since the resistance remains fixed, the output voltage decreases as the frequency is increased. The transfer functions formula for the low pass RL filters is $A = 1/1 + j(f_o/f_c)$. So, solve and plug the corner frequency and each of the operating frequencies into the formula to solve A, the phase angle, and the gain (A) in dB, as shown in conjunction with Figure 18-2.

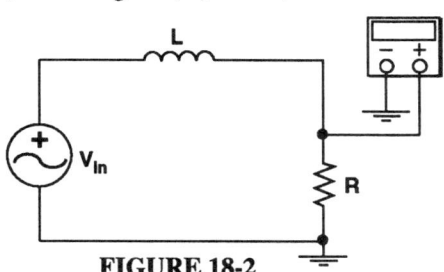

$$X_L = 2\pi fL$$

$$Z = \sqrt{R^2 + jX_L^2}$$

$$I = V_{in}/Z$$

$$V_R = IR$$

$$A(dB) = 20 \log V_R/V_{in}$$

$$f_C = \frac{R}{2\pi L}$$

$$A = \frac{1}{1 + j(f_o/f_c)}$$

$$\angle\theta = \tan^{-1}(f_o/f_c)$$

$$A(dB) = 20 \log A$$

FIGURE 18-2   **Standard Formulas**   **Transfer Function Formulas**

The instrument used on the EWB program to plot the response is the Bode Plotter shown in Figure 18-3(a). The expected frequency response of both the RC and RL low pass filter circuits is shown in Figure 18-3(b).

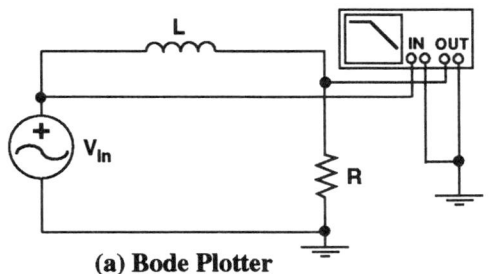 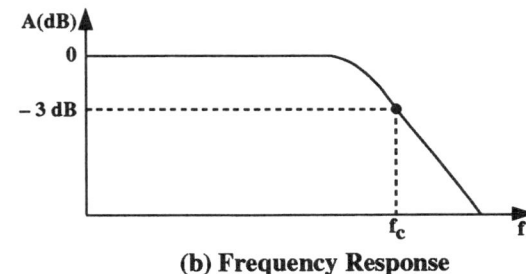

(a) Bode Plotter   (b) Frequency Response

FIGURE 18-3

LABORATORY EXERCISE

READING ASSIGNMENT: Basic Circuit Analysis for Electronics: 259-272

EXERCISE OBJECTIVES
To become familiar with :

- Transfer for low pass filters.
- Transfer functions of RC low pass filters.
- Transfer functions of RL low pass filters.

PROCEDURE

SECTION I: Low Pass RC Filters
PART A: Low Pass RC Filters
1. For the circuit connection of Figure 18-4(a), an input signal voltage of 3 V rms is applied to the circuit and the output signal is developed across the output capacitor C.

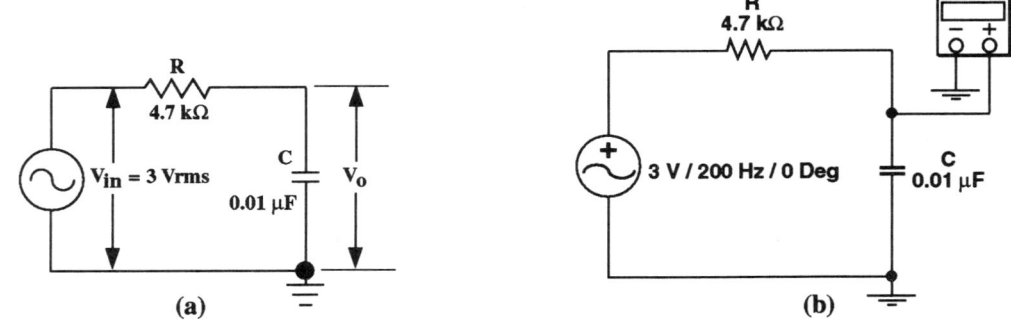

FIGURE 18-4

2. Calculate the output voltage across the output capacitor C at frequencies of 200 Hz, 1 kHz, 2 kHz, the corner frequency (f_c), 5 kHz, 10 kHz, and 50 kHz.
   a. Calculate the corner frequency, where $f_c = 1/2\pi RC$.
   b. Calculate the $A = V_o/V_{in}$ of the RC low pass filter for the frequencies 200 Hz, 1 kHz, 2 kHz, (f_c), 5 kHz, 10 kHz, and 50 kHz, where:

$$A = \frac{V_o}{V_{in}} = \frac{1}{1 + j(f_o/f_c)}$$

3. Convert the V_o/V_{in} to dB for the frequencies of 200 Hz, 1 kHz, 2 kHz, (f_c), 5 kHz, 10 kHz, and 50 kHz, where $A(dB) = 20 \log V_o/V_{in}$.

4. Insert the calculated values, as indicated, into Table 18-1.

PART B: Low Pass RC Filter Measurements
Open and select the circuit of Figure 18-4(b) from the File menu, or connect the circuit. Begin the RC low pass circuit measurements by applying an input signal voltage of 3 V rms at 200 Hz. The multimeter is set to ac (~) and volts (V).

Low Pass Filters — 129

1. Measure V_o, the output voltage across C, as shown in Figure 18-4(b).

2. Solve the V_o/V_{in} ratio at the frequency of 200 Hz and convert to decibels.

3. Switch the generator frequency to 1 kHz and make sure the input signal voltage is set to 3 V rms at 1 kHz.
   a. Measure the output voltage across the capacitor.
   b. Solve the V_o/V_{in} ratio at the frequency of 1 kHz and convert to decibels.

4. Repeat Step 3. Measure the output voltages at each of the remaining frequencies of 2 kHz, $f_c =$, 5 kHz, 10 kHz, and 50 kHz. Then, solve the V_o/V_{in} ratios at the various frequencies and convert to decibels.

5. Insert the measured output voltage values, as indicated, into Table 18-1.

TABLE 18-1	200 Hz	1 kHz	2 kHz	$f_c =$	5 kHz	10 kHz	50 kHz
CALC. V_o/V_{in}							
CALC. V_o/V_{in} (dB)							
MEASURE V_o							
SOLVED V_o/V_{in}							
SOLVED V_o/V_{in} (dB)							

PART C: Low Pass RC Bode Plot Measurements

Open and select the circuit of Figure 18-5 from the File menu, or connect the circuit.

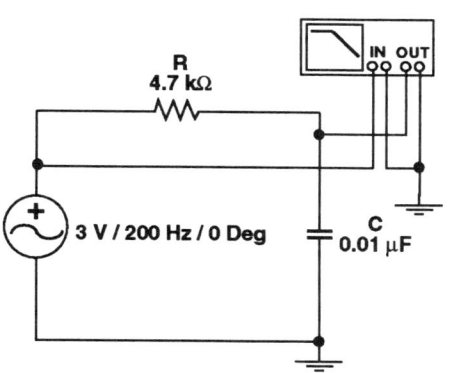

FIGURE 18-5

1. Set the initial (I) frequency to 1 Hz and the final (F) frequency to 1 MHz. Activate the circuit and monitor the response on the Bode plotter.
   a. Move the vertical cursor to 200 Hz and measure the output in decibels. The cursor can be dragged or incrementally moved by using the arrows located on the Bode plotter.
   b. Measure the dB level at each frequency, except at f_c, where both f_c and db are measured.

2. Repeat Step 1 measuring the output decibels at each of the remaining frequencies of 1 kHz, 2 kHz, the corner frequency (f_c), 5 kHz, 10 kHz, and 50 kHz.

3. Insert the measured output versus input voltage values in decibels, as indicated, into Table 18-2

TABLE 18-2	200 Hz	1 kHz	2 kHz	$f_c =$	5 kHz	10 kHz	50 kHz
MEASURED dB							

PART D: Plotting the Frequency Response

In the log-linear graph of Figure 18-6, use the derived values to plot the low pass frequency response curve. Then superimpose the measured voltmeter or Bode plotter A(dB) values and plot the frequency response. Label each response.

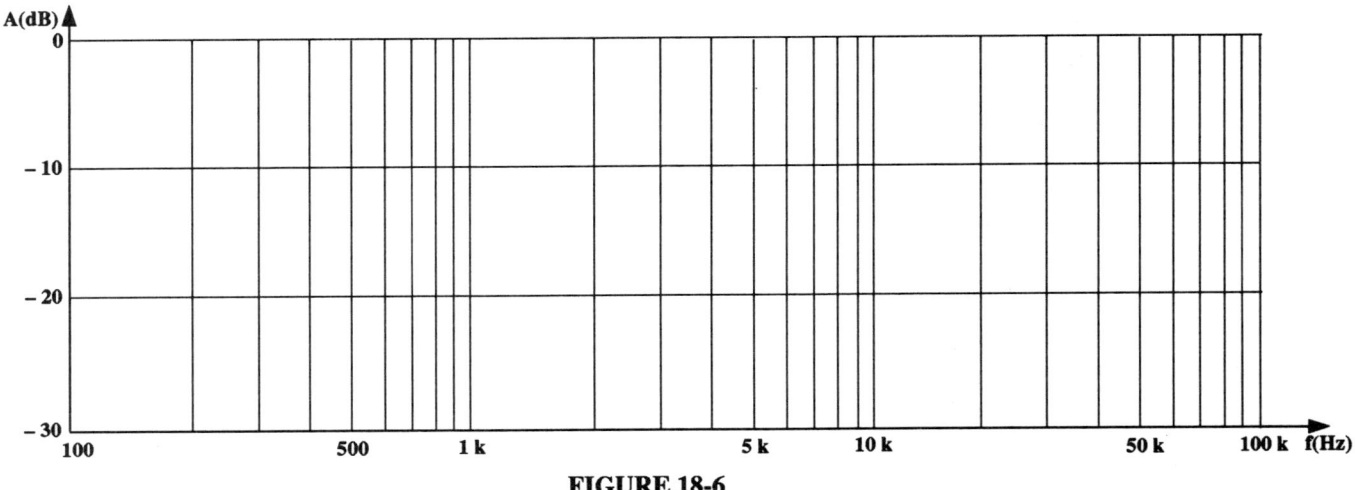

FIGURE 18-6

SECTION II: Low Pass RL Filters

PART A: Low Pass RL Circuit Calculations

For the circuit connection of Figure 18-7(a) an input signal of 3 V rms and 2 kHz (initially) is applied to the circuit. The output is developed across the 1.5 kΩ output resistor R.

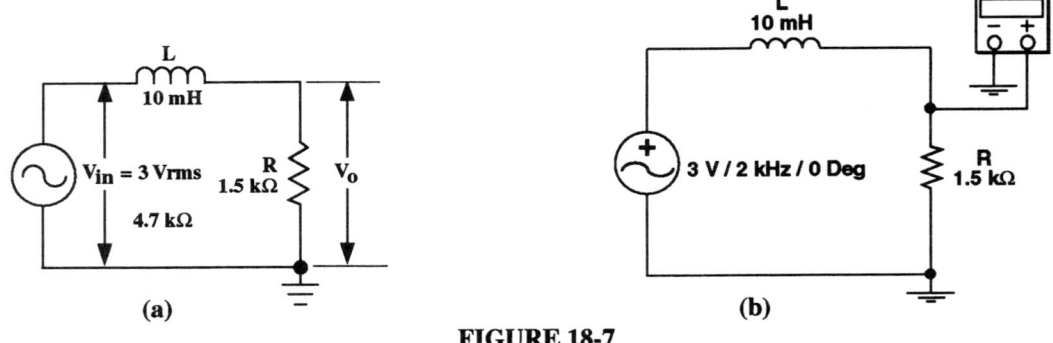

FIGURE 18-7

1. Calculate the output voltage across the output resistor R at frequencies of 2 kHz, 10 kHz, (f_c), 50 kHz, and 100 kHz.
   a. Calculate the corner frequency, where $f_c = R/2\pi L$.
   b. Calculate the $A = V_o/V_{in}$ of the RL low pass filter at 2 kHz, 10 kHz, the corner frequency (f_c), 50 kHz and 200 kHz, where:

$$A = \frac{V_o}{V_{in}} = \frac{1}{1 + j(f_o/f_c)}$$

2. Then, convert the V_o/V_{in} to dB at each of 2 kHz, 10 kHz, the corner frequency (f_c), 50 kHz, and 100 kHz, where $A(dB) = 20 \log V_o/V_{in}$.

3. Insert the calculated values, as indicated, into Table 18-3.

PART B: Low Pass RL Circuit Multimeter Measurements

Open and select the circuit of Figure 18-7(b) from the File menu, or connect the circuit. Begin the RL low pass circuit measurements by applying an input signal voltage of 3 V rms at 2 kHz.

Low Pass Filters — 131

1. Measure V_o, the output voltage across resistor R, as shown in Figure 18-7(b).

2. Solve the V_o/V_{in} ratio at the 2 kHz frequency and convert to decibels.

3. Switch the generator frequency to 10 kHz. Make sure the input signal voltage is set at 3 V rms at 10 kHz.
   a. Measure the output voltage across capacitor C.
   b. Solve the V_o/V_{in} ratio at the 10 kHz frequency and convert to decibels.

4. Repeat the measurement of the output voltage at each of the remaining frequencies of (fc), 50 kHz, and 200 kHz. Then, solve the V_o/V_{in} ratios of the various frequencies and convert to decibels.

5. Insert the measured output voltage values, as indicated, into Table 18-3

TABLE 18-3	2 kHz	10 kHz	f_c =	50 kHz	200 kHz
CALC. V_o/V_{in}					
CALC. V_o/V_{in} (dB)					
MEASURED V_o					
SOLVED V_o/V_{in}					
SOLVED V_o/V_{in} (dB)					

PART C: Plotting the Frequency Response

Open and select the circuit of Figure 18-8 from the File menu, or connect the circuit. Set the low initial (I) frequency to 1 Hz and the final (F) frequency to 1 MHz.

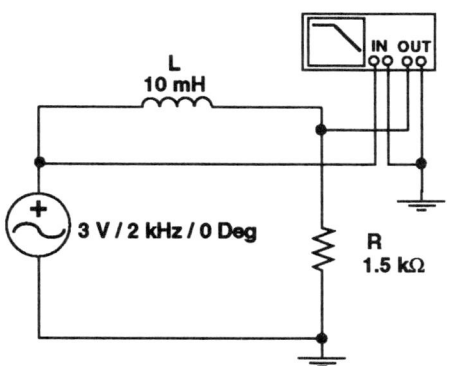

FIGURE 18-8

1. Activate the circuit and monitor the response on the Bode plotter.
   a. Move the vertical cursor to 2 kHz and measure the output voltage in decibels. The cursor can be dragged and incrementally moved by using the arrows located on the Bode plotter.
   b. Measure both the frequency and dB level, which are directly indicated on the Bode plotter.

2. Repeat Step 1, measuring the decibels at the remaining frequencies of 10 kHz, f_c, 50 kHz and 200 kHz.

3. Insert the measured output versus input voltages in decibels, as indicated, into Table 18-4.

TABLE 18-4	2 kHz	10 kHz	f_c =	50 kHz	200 kHz
MEASURED					

132 — Basic Circuit Analysis For Electronics Using Electronic Workbench®

PART D: Plotting the Frequency Response

In the log-linear graph of Figure 18-9, use the derived values to plot the low pass frequency response curve, and then superimpose the measured voltmeter or Bode plotter A(dB) values and plot the frequency response. Label each response.

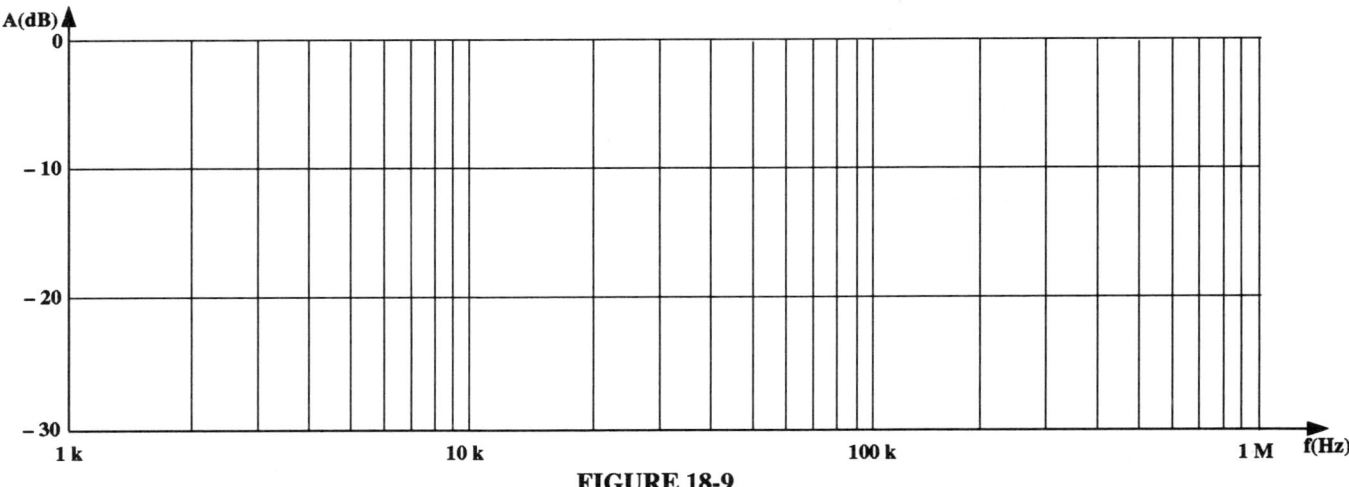

FIGURE 18-9

Questions and Problems: Basic Circuit Analysis for Electronics: 276-277

CHAPTER 19
HIGH PASS FILTERS

INTRODUCTION

High pass filters provide a frequency response opposite to low pass filters. In the basic two component RC high pass filter circuit of Figure 19-1, the capacitive reactance decreases as the frequency is increased, but the resistance remains fixed. So as the frequency is increased the output voltage increases. The standard formulas and the transfer function formulas for solving the output voltage at any given frequency are shown in conjunction with Figure 19-1. The advantage of transfer functions is that the formula is easily applied. For the RC high pass filter the formula (which is the same for RL circuits) is $A = j(f_o/f_c)/1 + j(f_o/f_c)$. So, begin by solving the corner frequency, and then plug both the corner frequency and each of the operating frequencies into the formula to solve A, the phase angle, and the gain (A) in dB.

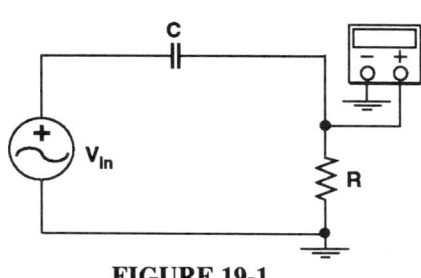

$$X_C = \frac{1}{2\pi fC}$$
$$Z = \sqrt{R^2 + jX_C^2}$$
$$V_R = IR$$
$$V_C = IX_C$$
$$A(dB) = 20 \log V_R/V_{in}$$

$$f_c = \frac{1}{2\pi RC}$$
$$A = \frac{j(f_o/f_c)}{1 + j(f_o/f_c)}$$
$$\angle\theta = \tan^{-1}(f_o/f_c)$$
$$A(dB) = 20 \log A$$

FIGURE 19-1 **Standard Formulas** **Transfer Function Formulas**

In the basic two component RL circuit of Figure 19-2, the resistance remains fixed, but the inductive reactance of the inductor increases as the frequency is increased. So, the output voltage increases as the frequency is increased. The transfer functions formula for the low pass RL is $A = j(f_o/f_c)/1 + j(f_o/f_c)$. So, solve and plug the corner frequency and each of the operating frequencies into the formula to solve A, the phase angle and then the gain (A) in dB.

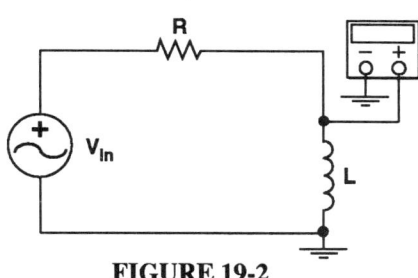

$$X_L = 2\pi fL$$
$$Z = \sqrt{R^2 + jX_L^2}$$
$$I = V_{in}/Z$$
$$V_R = IX_L$$
$$A(dB) = 20 \log V_{XL}/V_{in}$$

$$f_c = \frac{R}{2\pi L}$$
$$A = \frac{j(f_o/f_c)}{1 + j(f_o/f_c)}$$
$$\angle\theta = \tan^{-1}(f_o/f_c)$$
$$A(dB) = 20 \log A$$

FIGURE 19-2 **Standard Formulas** **Transfer Function Formulas**

The instrument used on the EWB program to plot the response is the Bode Plotter shown in figure 19-3(a). The expected frequency response of both the RC and RL high pass filter connections is shown in Figure 19-3(b).

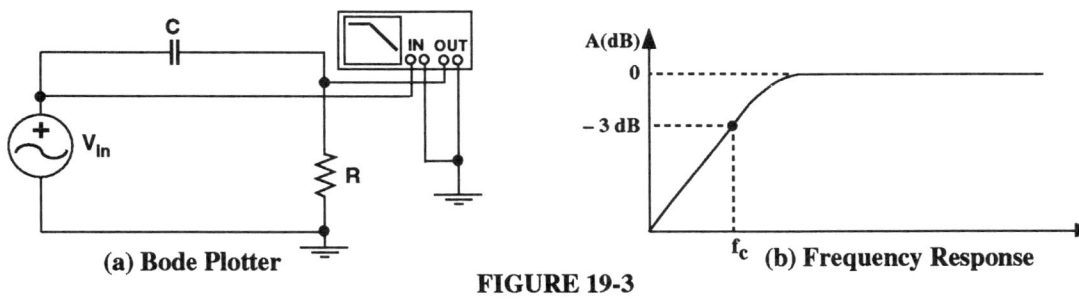

(a) Bode Plotter   (b) Frequency Response

FIGURE 19-3

LABORATORY EXERCISE

READING ASSIGNMENT: Basic Circuit Analysis for Electronics: 278-290

EXERCISE OBJECTIVES
To become familiar with :

- Transfer for high pass filters.
- Transfer functions of RC high pass filters.
- Transfer functions of RL high pass filters.

PROCEDURE

SECTION I: High Pass RC Filters
PART A: Low Pass RC Filters

1. For the circuit connection of Figure 19-4(a), an input signal voltage of 3 V rms is applied to the circuit and the output signal is developed across the output resistor R.

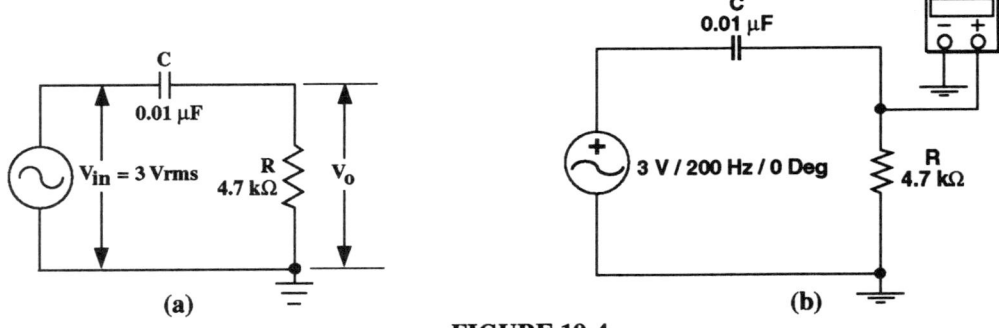

FIGURE 19-4

2. Calculate the output voltage across the output resistor R at frequencies of 200 Hz, 1 kHz, 2 kHz, the corner frequency (f_c), 5 kHz, 10 kHz, and 50 kHz.
   a. Calculate the corner frequency, where $f_c = 1/2\pi RC$.
   b. Calculate the $A = V_o/V_{in}$ of the RC low pass filter for the frequencies 200 Hz, 1 kHz, 2 kHz, (fc), 5 kHz, 10 kHz, and 50 kHz, where:

$$A = \frac{V_o}{V_{in}} = \frac{j(f_o/f_c)}{1 + j(f_o/f_c)}$$

3. Convert the V_o/V_{in} to dB for the frequencies of 200 Hz, 1 kHz, 2 kHz, (f_c), 5 kHz, 10 kHz, and 50 kHz, where $A(dB) = 20 \log V_o/V_{in}$.

4. Insert the calculated values, as indicated, into Table 19-1.

PART B: High Pass RC Filter Measurements
Open and select the circuit of Figure 19-4(b) from the File menu, or connect the circuit. Begin the RC high pass circuit measurements by applying an input signal voltage of 3 V rms at 200 Hz. The multimeter is set to ac (~) and volts (V).

High Pass Filters 135

1. Measure V_o, the output voltage across resistor R, as shown in Figure 19-4(b).

2. Solve the V_o/V_{in} ratio at the frequency of 200 Hz and convert to decibels.

3. Switch the generator frequency to 1 kHz and make sure the input signal voltage is set to 3 V rms at 1 kHz.
    a. Measure the output voltage across the resistor R.
    b. Solve the V_o/V_{in} ratio at the frequency of 1 kHz and convert to decibels.

4. Repeat the measurement of the output voltages at each of the remaining frequencies of 2 kHz, (f_c), 5 kHz, 10 kHz, and 50 kHz. Then, solve the V_o/V_{in} ratios at the various frequencies and convert to decibels.

5. Insert the measured output voltage values, as indicated, into Table 19-1.

TABLE 19-1	200 Hz	1 kHz	2 kHz	$f_c =$	5 kHz	10 kHz	50 kHz
CALC. V_o/V_{in}							
CALC. V_o/V_{in} (dB)							
MEASURED V_o							
SOLVED V_o/V_{in}							
SOLVED V_o/V_{in} (dB)							

PART C: High Pass RC Bode Plot Measurements
Open and select the circuit of Figure 19-5 from the File menu, or connect the circuit.

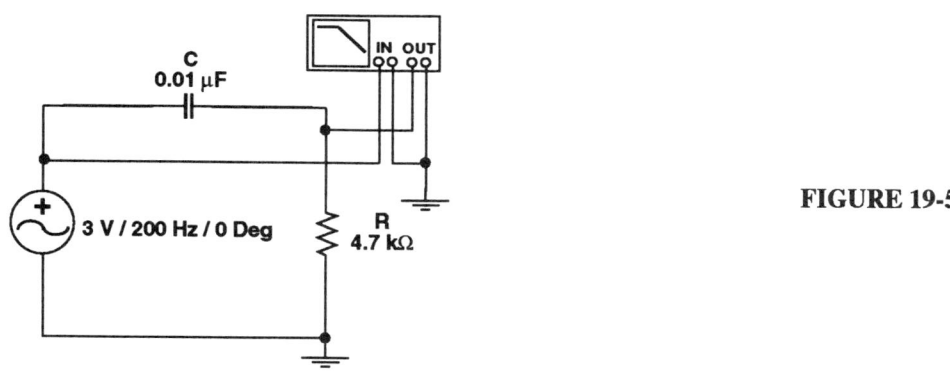

FIGURE 19-5

1. Set the initial (I) frequency to 1 Hz and the final (f) frequency to 1 MHz. Activate the circuit and monitor the response on the Bode plotter.
    a. Move the vertical cursor to 200 Hz and measure the output in decibels. The cursor can be dragged or incrementally moved by using the arrows located on the Bode plotter.
    b. Measure both the frequency and the dB level, which are directly indicated on the Bode plotter.

2. Repeat Step 1 measuring the output decibels at each of the remaining frequencies of 1 kHz, 2 kHz, the corner frequency (f_c), 5 kHz, 10 kHz, and 50 kHz.

3. Insert the measured output versus input voltage values in decibels, as indicated, into Table 19-2

TABLE 19-2	200 Hz	1 kHz	2 kHz	$f_c =$	5 kHz	10 kHz	50 kHz
MEASURED							

PART D: Plotting the Frequency Response

In the log-linear graph of Figure 19-6, use the derived values to plot the high pass frequency response curve, and then superimpose the measured voltmeter or Bode plotter A(dB) values and plot the frequency response curve. Label each response.

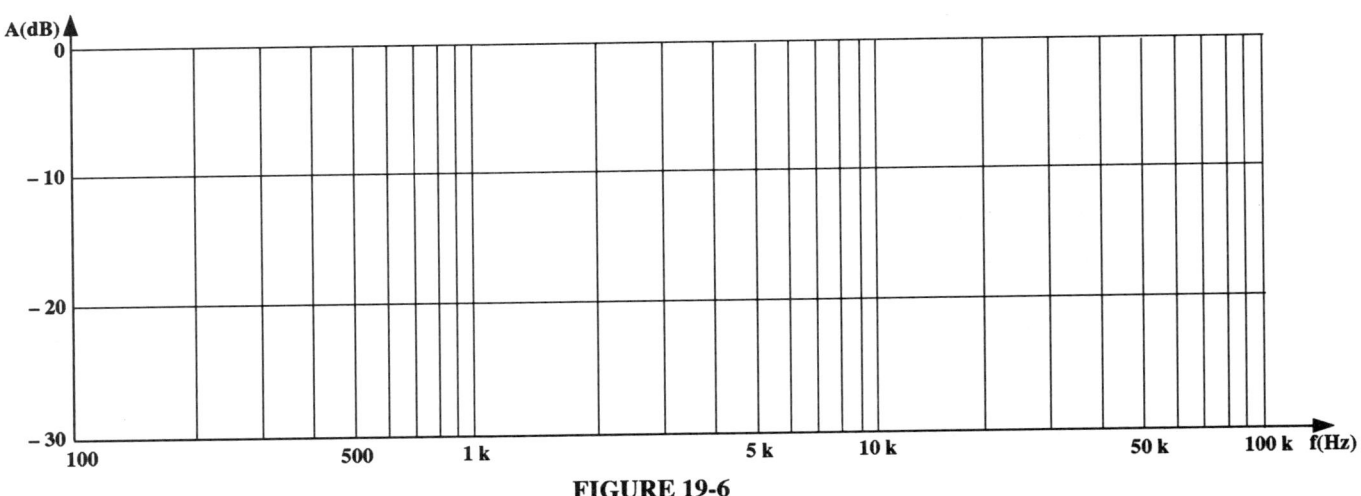

FIGURE 19-6

SECTION II: High Pass RL Filters
PART A: High Pass RL Circuit Calculations

For the circuit connection of Figure 19-7(a) and input signal of 3 V rms at 2 kHz (initially) are applied to the circuit. The output is developed across the 1.5 kΩ output inductor L.

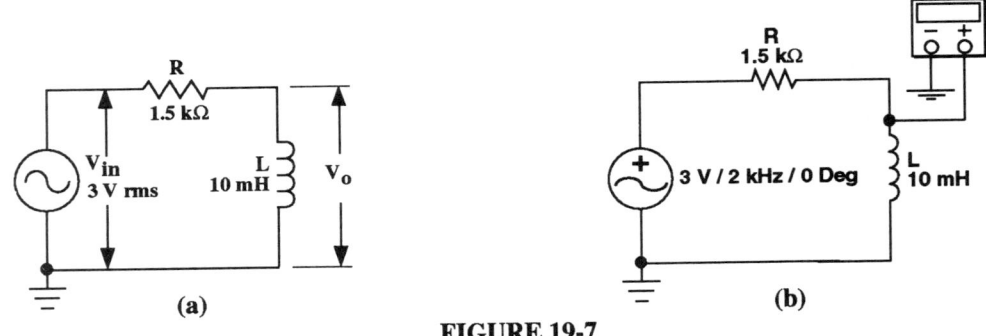

FIGURE 19-7

1. Calculate the output voltage across the output inductor L at frequencies of 2 kHz, 5 kHz, 10 kHz, the corner frequency (f_c), 50 kHz, and 500 kHz.
    a. Calculate the corner frequency, where $f_c = R/2\pi L$.
    b. Calculate the $A = V_o/V_{in}$ of the RL low pass filter at 2 kHz, 10 kHz, the corner frequency (f_c), 50 kHz, and 200 kHz where:

$$A = \frac{V_o}{V_{in}} = \frac{j(f_o/f_c)}{1 + j(f_o/f_c)}$$

2. Then, convert the V_o/V_{in} to dB at each of 2 kHz, 10 kHz, the corner frequency (f_c), 50 kHz, and 200 kHz: where: $A(dB) = 20 \log V_o/V_{in}$.

3. Insert the calculated values, as indicated, into Table 19-2.

PART B: Low Pass RL Circuit Multimeter Measurements

Open and select the circuit of Figure 19-7(b) from the File menu, or connect the circuit. Begin the RL High pass circuit measurements by applying an input signal voltage of 3 V rms at 2 kHz.

1. Measure V_o, the output voltage across inductor L, as shown in Figure 19-7(b).

2. Solve the V_o/V_{in} ratio at the 2 kHz frequency and convert to decibels.

3. Switch the generator frequency to 10 kHz. Make sure the input signal voltage is set at 3 V rms at 10 kHz.
   a. Measure the output voltage across inductor L.
   b. Solve the V_o/V_{in} ratio at the 10 kHz frequency and convert to decibels.

4. Repeat the measurements of the output voltage at each of the remaining frequencies of (f_c), 50 kHz, and 200 kHz. Then solve the V_o/V_{in} ratios of the various frequencies and convert to decibels.

5. Insert the measured output voltage values, as indicated, into Table 19-3

TABLE 19-3	2 kHz	10 kHz	$f_c =$	50 kHz	200 kHz
CALC. V_o/V_{in}					
CALC. V_o/V_{in} (dB)					
MEASURED V_o					
SOLVED V_o/V_{in}					
SOLVED V_o/V_{in} (dB)					

PART C: Plotting the Frequency Response

Open and select the circuit of Figure 19-8 from the File menu, or connect the circuit. Set the low initial (I) frequency to 1 Hz and the final (F) frequency to 1 MHz.

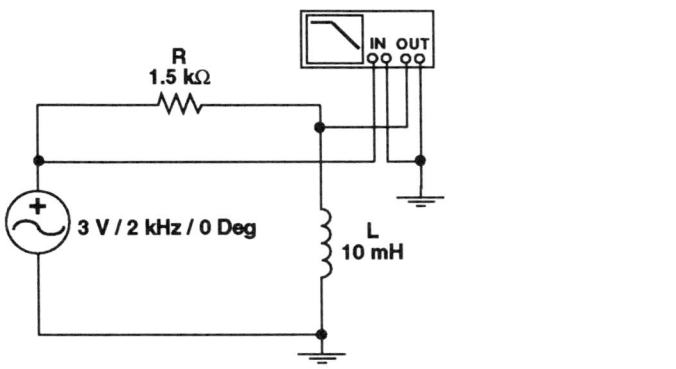

FIGURE 19-8

1. Activate the circuit and monitor the response on the Bode plotter.
   a. Move the vertical cursor to 2 kHz and measure the output voltage in decibels. The cursor can be dragged and incrementally moved by using the arrows located on the Bode plotter.
   b. Measure both the frequency and dB level, which are directly indicated on the Bode plotter.

2. Repeat Step 1, measuring the decibels at the remaining frequencies of 10 kHz, f_c, 50 kHz and 200 kHz.

3. Insert the measured output versus input voltages in decibels, as indicated, into Table 19-4.

TABLE 19-4	2 kHz	10 kHz	$f_c =$	50 kHz	200 kHz
MEASURED					

PART D: Plotting the Frequency Response

In the log-linear graph of Figure 19-9, use the derived values to plot the high pass frequency response curve, and then superimpose the measured voltmeter or Bode plotter A(dB) values and plot the frequency response curve. Label each response.

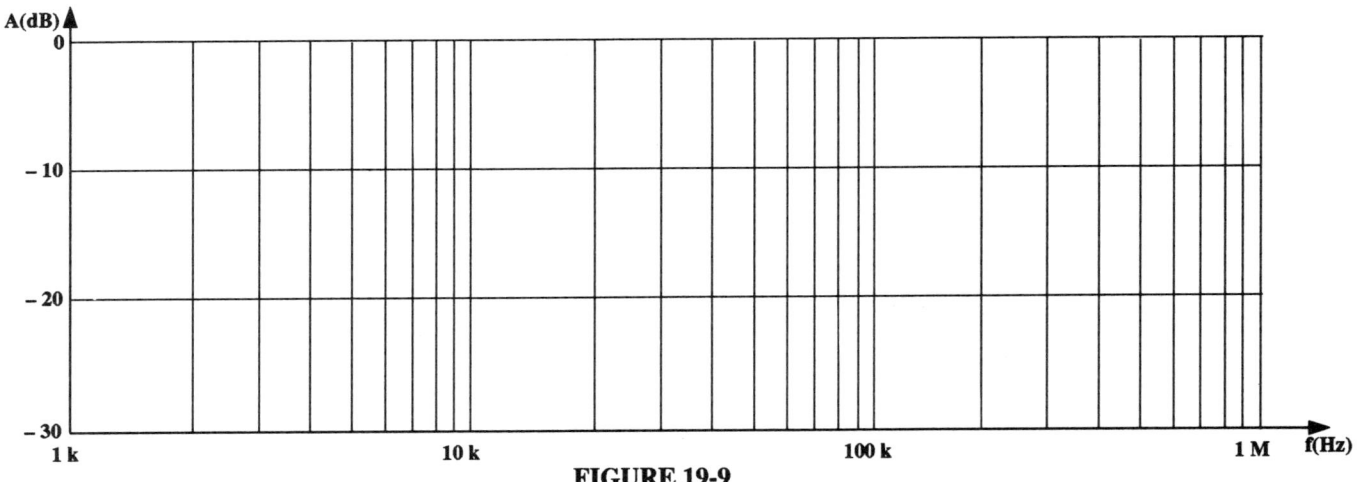

FIGURE 19-9

Questions and Problems: Basic Circuit Analysis for Electronics: 294-295

CHAPTER 20
BANDPASS RC FILTERS

INTRODUCTION
Bandpass filters are used to select a band of frequencies and reject all others. RC bandpass filters are normally used at the lower frequencies, typically below 100 kHz, and are generally wideband. In the laboratory exercise two RC bandpass circuit configurations will be analyzed, and although the two circuits are identical, the component values of C1 and C2 in the circuits are different and so is the analysis.

WIDEBAND RC BANDPASS FILTER
In the circuit of Figure 20-1 (a), where R1 equals R2 but the C1 value is 100 times greater than C2, so the low and high corner frequencies are widely separated. At mid-band frequencies (f_o), the response is effectively flat, as shown in Figure 20-2(b), and provides the maximum output signal voltage possible (V_o) for the circuit. Then, at the low (f_L) and high corner (f_H) frequencies the signal voltage drops to 0.707 (− 3 dB) of the mid-band frequency amplitude. The mid-band frequency formula, the output voltage at the midband frequency, the low and high corner frequencies, the gain (A) taken at any frequency and its value in decibels is shown in association with figure 20-1(b).

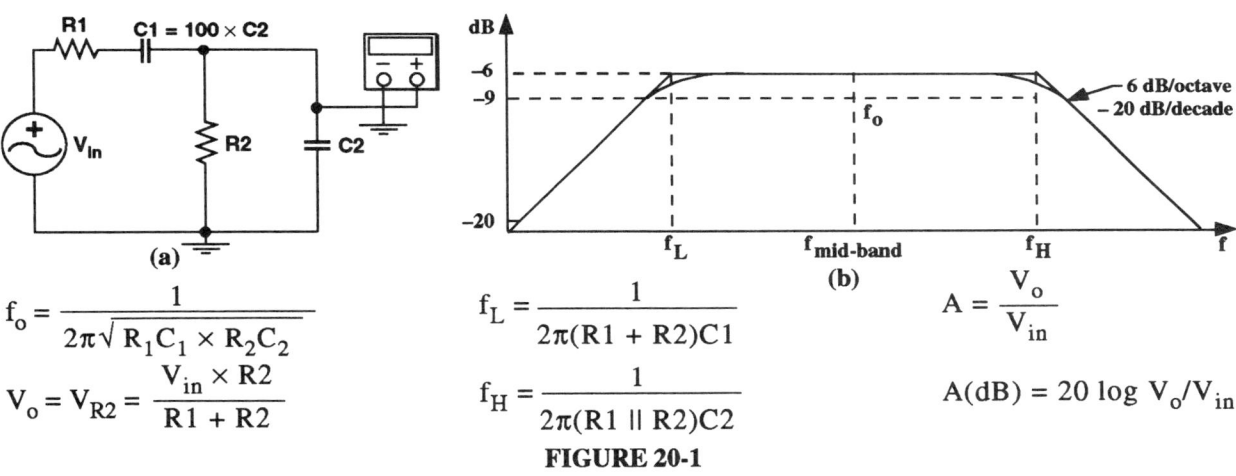

$$f_o = \frac{1}{2\pi\sqrt{R_1C_1 \times R_2C_2}}$$

$$V_o = V_{R2} = \frac{V_{in} \times R2}{R1 + R2}$$

$$f_L = \frac{1}{2\pi(R1 + R2)C1}$$

$$f_H = \frac{1}{2\pi(R1 \parallel R2)C2}$$

$$A = \frac{V_o}{V_{in}}$$

$$A(dB) = 20 \log V_o/V_{in}$$

FIGURE 20-1

NARROWBAND RC BANDPASS FILTER
In the circuit of Figure 20-2, R = R1 = R2 and C = C1 = C2, so the bandwidth is narrower and the analysis is different. The mid-band frequency formula, the output voltage at the midband frequency, the low and high corner frequencies, the gain (A) taken at any frequency and its value in decibels are shown in conjunction with the circuit of Figure 20-2

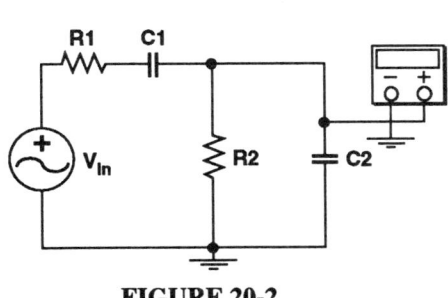

FIGURE 20-2

$$f_o = \frac{1}{2\pi RC}$$

$$V_o = \frac{V_{in} \times Z_p}{Z_s + Z_p}$$

where: $Z_s = R1 - jX_{C1}$

$Z_p = R2 \parallel -jX_{C2}$

$f_L = f_o/3.3$

$f_H = f_o \times 3.3$

$$A = \frac{V_o}{V_{in}}$$

$$A(dB) = 20 \log V_o/V_{in}$$

140 — Basic Circuit Analysis For Electronics Using Electronic Workbench®

LABORATORY EXERCISE

READING ASSIGNMENT: Basic Circuit Analysis for Electronics: 296-302
EXERCISE OBJECTIVES
To become familiar with:

- Broadband, bandpass, RC filters.
- R1 = R2, C1 = C2, bandpass filter.

PROCEDURE

SECTION I: Broadband, Bandpass, RC Filter
PART A: Pre-laboratory Calculations

1. Analyze the RC band pass circuit of Figure 20-3(a) at the mid-band frequency and at the low and high corner frequencies. Use a V_{in} of 6 Vp-p.

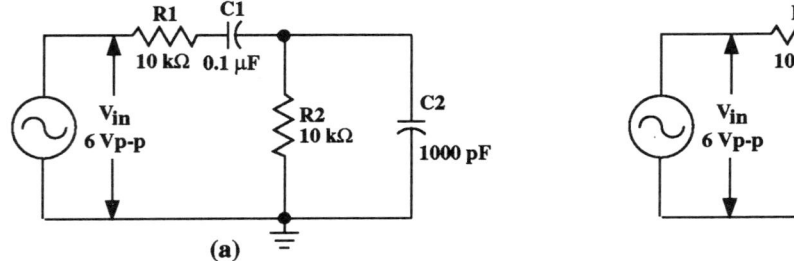

FIGURE 20-3

2. Mid-band frequency calculations: At the mid-band frequency the equivalent circuit approximates R1 and R2 in series, as shown in Figure 20-3(b). At the mid-band frequency, the capacitive reactance of C1 and C2 can be safely ignored in the calculations.

   a. Calculate the mid-band operating frequency where: $f_o = \dfrac{1}{2\pi \sqrt{R_1 C_1 \times R_2 C_2}}$

   b. Calculate the output voltage across the R2 resistor where: $V_o \approx \dfrac{V_{in} \times R2}{R1 + R2}$.

   c. Convert the V_o/V_{in} ratio to decibels where: $A(dB) = 20 \log V_o/V_{in}$.

3. Low corner frequency calculations: At the low corner frequency the capacitive reactance X_{C1} equals the series circuit resistance of R1 + R2 as shown in the equivalent circuit of Figure 20-4(a). At the low corner frequency X_{C2} can be safely ignored in the calculations.

   a. Calculate the low corner frequency where: $f_L = \dfrac{1}{2\pi(R_1 + R_2)C_1}$.

   b. Calculate the capacitive reactance X_{C1} where: $X_{C1} = 1/(2\pi f_L C_1)$.

   c. Calculate the output voltage across resistor R2 at the low corner frequency where: $V_o \approx \dfrac{V_{in} \times R2}{\sqrt{2}\,(R1 + R2)}$

   d. Then convert the V_o/V_{in} ratio to decibels where: $A(dB) = 20 \log V_o/V_{in}$

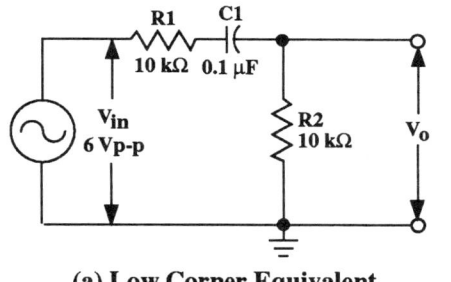
(a) Low Corner Equivalent

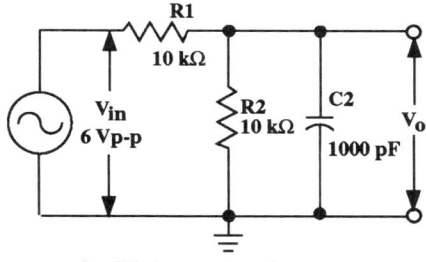
(b) High Corner Equivalent

FIGURE 20-4

4. High Corner Frequency calculations: At the high corner frequency, the capacitive reactance X_{C2} equals the R_{TH} of the circuit (R1 ∥ R2) as shown in the equivalent circuit of Figure 20-4(b). At the high corner frequency X_{C1} can be safely ignored in the calculations.

   a. Calculate the high corner frequency where: $f_H = \dfrac{1}{2\pi(R_1 \| R_2)C_2}$.

   b. Calculate the capacitive reactance of X_{C2} where: $X_{C2} = 1/(2\pi f_H C_2)$.

   c. Calculate the output voltage across the R2 resistor, where: $V_o \approx \dfrac{V_{in} \times R2}{\sqrt{2}\,(R1 + R2)}$.

   d. Then convert the V_o/V_{in} ratio to decibels where: $A(dB) = 20 \log V_o/V_{in}$.

5. Insert the calculated values, as indicated, into Table 20-1.

PART B: Circuit Measurements

Open and select the circuit of Figure 20-5 from the File menu, or connect the circuit. Set the ac source at the mid-band frequency and 6 Vp-p. Set the the oscilloscope at 1 V/div and 0.1 ms/div

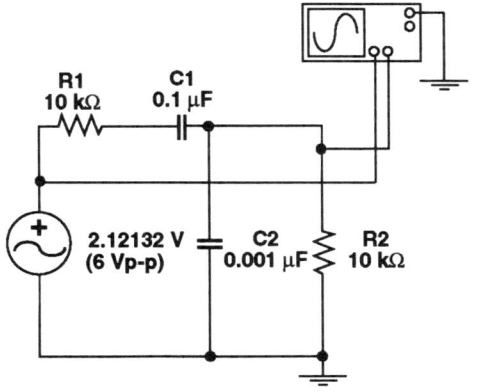

FIGURE 20-5

1. Measure the output voltage at the mid-band, low, and high corner frequencies.
   a. Mid-band frequency conditions
      1. Use an input voltage of 6 Vp-p with the frequency set to the calculated mid-band frequency.
      2. Measure the output voltage and convert the V_o/V_{in} ratio to decibels, where $A(dB) = 20 \log V_o/V_{in}$.
   b. Low corner frequency conditions
      1. Vary the frequency to the calculated low corner frequency.
      2. Measure the output voltage at the calculated low corner frequency.
      3. Convert the V_o/V_{in} ratio to decibels where: $A(dB) = 20 \log V_o/V_{in}$.
   c. High corner frequency conditions
      1. Vary the frequency to the calculated high corner frequency.
      2. Measure the output voltage at the calculated high corner frequency.
      3. Then convert the V_o/V_{in} ratio to decibels where: $A(dB) = 20 \log V_o/V_{in}$.

2. Insert the measured values, as indicated, into Table 20-1.

TABLE 20-1	Midband Frequency			Low Corner Frequency				High Corner Frequency			
	f_o	V_o	A(dB)	f_L	X_{C1}	V_o	A(dB)	f_H	X_{C2}	V_o	A(dB)
CALCULATED											
MEASURED											

PART C: Bandpass RC Circuit Bode Plotter Measurements

Open and select the circuit of Figure 20-6 from the File menu, or connect the circuit. Set the Initial frequency (I) to 1 Hz and the Final frequency (F) to 1 MHz.

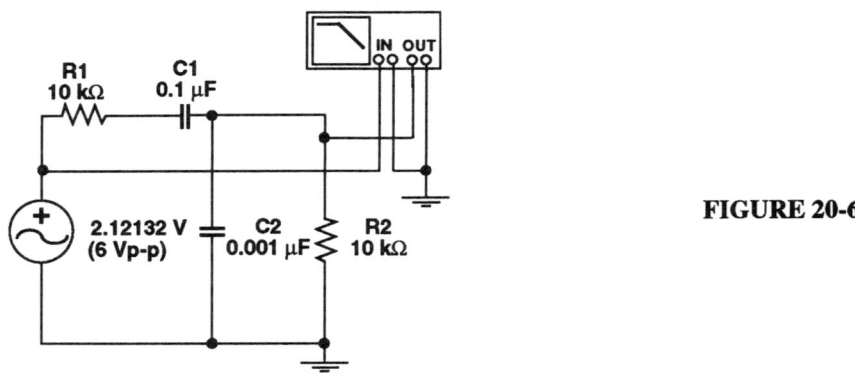

FIGURE 20-6

1. Activate the circuit and monitor the response on the Bode plotter.
   a. Move the vertical cursor the mid-band frequency and measure the output voltage in decibels. The cursor can be dragged or incrementally moved by using the arrows located on the plotter.
   b. Measure both the frequency and db level. Both values are directly indicated on the Bode plotter.

2. Repeat Step 1, measuring the output in decibels at the low corner and at the high corner frequencies. Then, continue the measurements at 0.5 f_L and 2 f_L and at 0.5 f_H and 2 f_H.

3. Insert the measured output voltage values in decibels, as indicated, into Table 20-2.

TABLE 20-2	0.5 f_L	f_L	2 f_L	f_o(mid-band)	0.5 f_H	f_H	2 f_H
Frequency (Calc.)							
Frequency (Meas.)							
A(dB) (Meas.)							

PART D: Plotting the Frequency Response

On the log-linear graph of Figure 20-7, use both the calculated and measured values to plot the bandpass response curve, and then superimpose the measured voltmeter or Bode plotter A(dB) values to plot the response curve. Label each response.

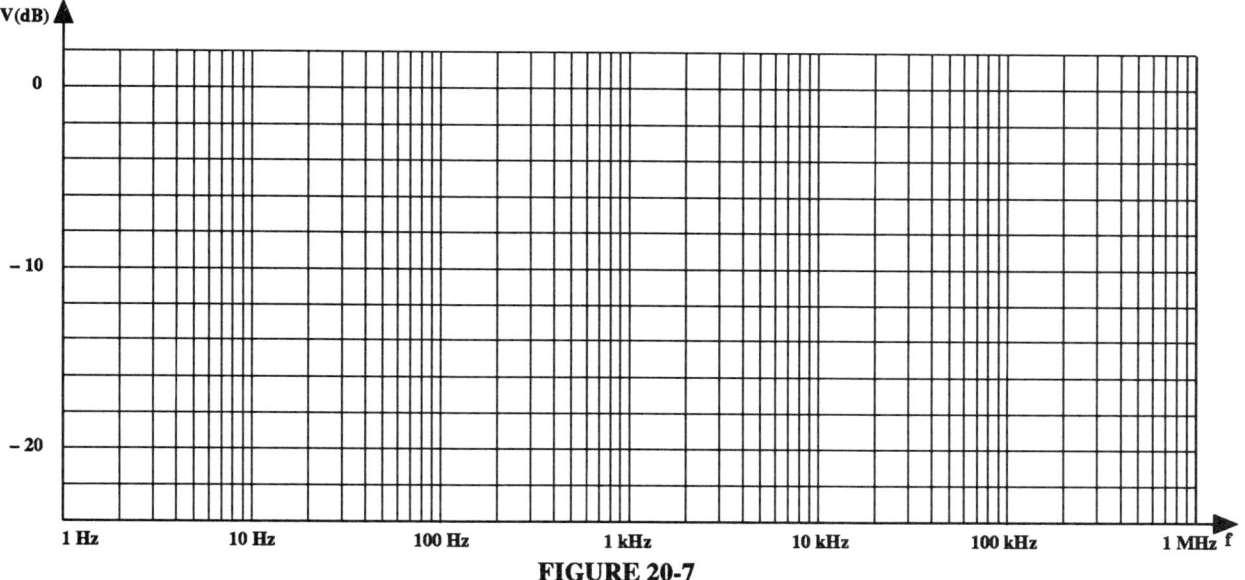

FIGURE 20-7

SECTION II: Series-Parallel, R1 = R2, C1 = C2 Circuit
PART A: Pre-laboratory Calculations
1. Analyze the RC band pass circuit of Figure 20-8(a) at the mid-band frequency and then at the low and high corner frequencies. Use a V_{in} of 6 Vp-p.

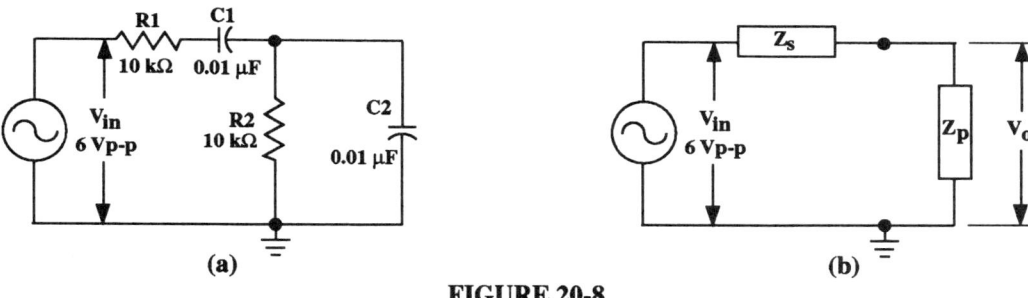

FIGURE 20-8

2. Mid-band frequency calculations: At the mid-band frequency the equivalent circuit approximates Z_s and Z_p in series, as shown in Figure 20-8(b).

   a. Calculate the mid-band operating frequency where: $f_o = \dfrac{1}{2\pi\sqrt{R_1C_1 \times R_2C_2}} = \dfrac{1}{2\pi R_1C_1}$.

   b. Calculate the X_{C1} and then the series $Z_s = R1 - jX_{C1}$ value and the associated phase angle.

   c. Then calculate the X_{C2} and then the series $Z_p = R2 \parallel -jX_{C2}$ value and the associated phase angle.

   d. Calculate the output voltage across the R2 resistor where: $V_o \approx \dfrac{V_{in} \times Z_p\angle\theta}{Z_s\angle\theta + Z_p\angle\theta}$.

   e. Then convert the V_o/V_{in} ratio to decibels, where $A(dB) = 20 \log V_o/V_{in}$.

NOTE: Since R1 = R2 and C1 = C2 in the circuit, the mid-band frequency can be solved using either equation $f_o = 1/2\pi R_1C_1$ or $f_o = 1/2\pi R_2C_2$.

3. Low and high corner frequency calculations: The low corner frequency can be approximated by dividing the calculated mid-band frequency by 3.3, while the high corner frequency can be approximated by multiplying the calculated mid-band frequency by 3.3.

a. Calculate the approximate low corner frequency where: $f_L = f_o/3.3$.
b. Calculate the approximate high corner frequency, where; $f_H = f_o \times 3.3$.

PART B: Circuit Measurements

1. Open and select the circuit of Figure 20-9 from the File menu, or connect the circuit. Use an input voltage of 6 Vp-p with the scope set to 1 V/div and 0.1 ms/div.

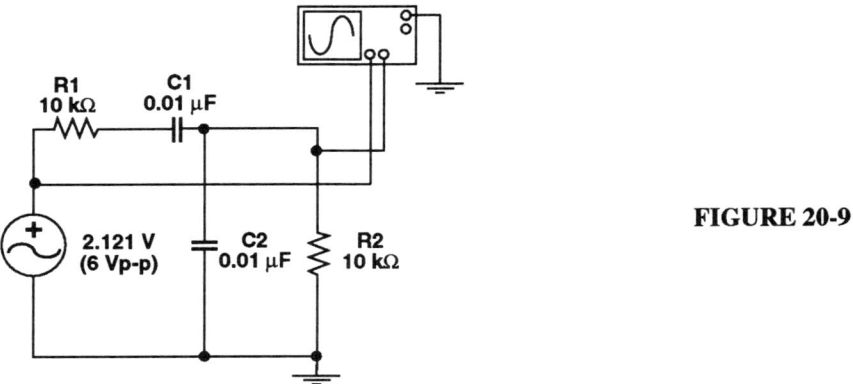

FIGURE 20-9

2. Measure the output voltage at the mid-band, low, and high corner frequencies.
   a. Mid-band frequency conditions
      1. Use an input voltage of 6 Vp-p with the frequency set to the calculated mid-band frequency.
      2. Measure the output voltage and convert the V_o/V_{in} ratio to decibels, where $A(dB) = 20 \log V_o/V_{in}$.
   b. Low corner frequency conditions
      1. Vary the frequency to the calculated low corner frequency
      2. Measure the output voltage at the calculated low corner frequency.
      3. Convert the V_o/V_{in} ratio to decibels where: $A(dB) = 20 \log V_o/V_{in}$.
   c. High corner frequency conditions
      1. Vary the frequency to the calculated high corner frequency.
      2. Measure the output voltage at the calculated high corner frequency.
      3. Then convert the V_o/V_{in} ratio to decibels where: $A(dB) = 20 \log V_o/V_{in}$.

3. Insert the calculated and measured values, as indicated, into Table 20-3.

TABLE 20-3	Midband Frequency					Low Corner Frequency			High Corner Frequency		
	f_o	Z_s	Z_p	V_o	A(dB)	f_L	V_o	A(dB)	f_H	V_o	A(dB)
CALCULATED											
MEASURED											

PART C: Bandpass RC Circuit Bode Plotter Measurements

Open and select the circuit of Figure 20-10 from the File menu, or connect the circuit. Set the initial frequency (I) to 1 Hz and the final frequency (F) to 1 MHz and 0.0 dB.

1. Activate the circuit and monitor the response on the Bode plotter.
   a. Move the vertical cursor the mid-band frequency and measure the output voltage in decibels. The cursor can be dragged or incrementally moved by using the arrows located on the plotter.
   b. Measure both the frequency and db level. Both values are directly indicated on the Bode plotter.

2. Repeat Step 1, measuring the output in decibels at the low corner and at the high corner frequencies. Then, continue the measurements at 0.5 f_L and 2 f_L and at 0.5 f_H and 2 f_H.

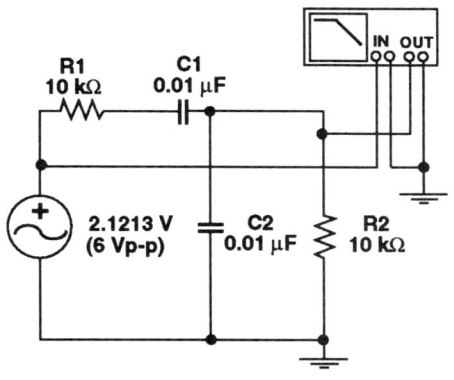

FIGURE 20-10

3. Insert the measured output voltage values in decibels, as indicated, into Table 20-4.

TABLE 20-4	$0.5 f_L$	f_L	$2 f_L$	f_o(mid-band)	$0.5 f_H$	f_H	$2 f_H$
Frequency (Calc.)							
Frequency (Meas)							
A(dB) (Meas.)							

PART D: Plotting the Frequency Response

On the log-linear graph of Figure 20-11, use both the calculated and measured values to plot the bandpass response curve, and then superimpose the measured voltmeter or Bode plotter A(dB) values to plot the response curve. Label each response.

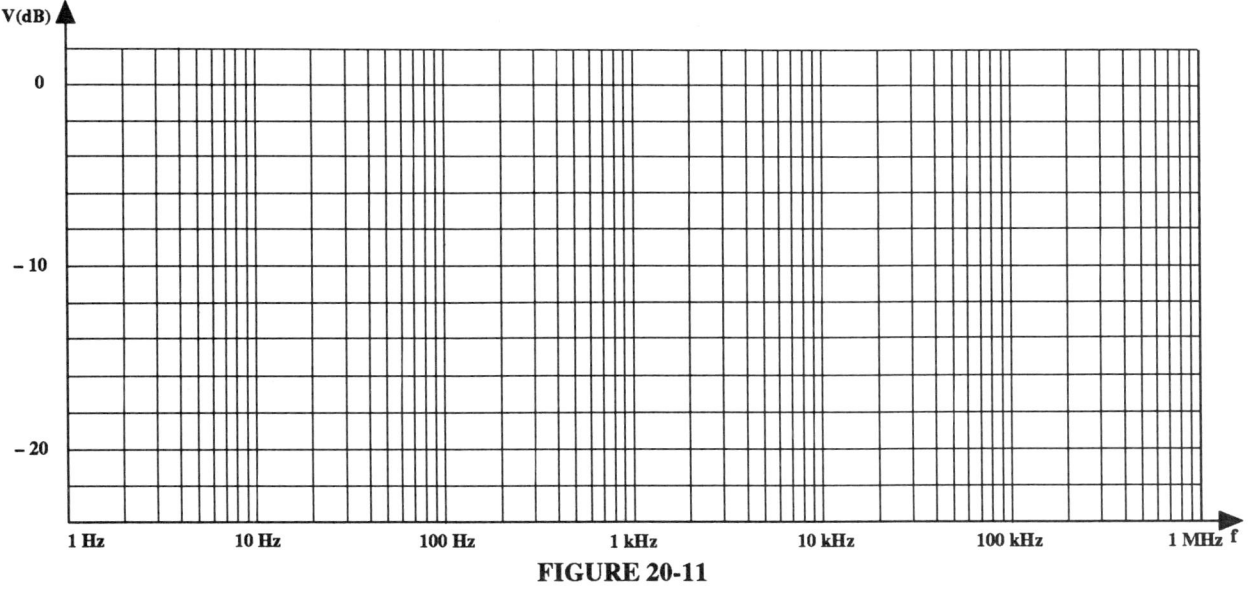

FIGURE 20-11

Questions and Problems: Basic Circuit Analysis for Electronics: 307-308

CHAPTER 21
BANDPASS AND BAND REJECT LC FILTERS

INTRODUCTION
Bandpass filters are used to select a narrow band of frequencies and are either series or parallel resonant circuits constructed from LC components. The frequency at which resonance occurs (f_r) is derived, knowing that $X_L = X_C$, $X_L = 2\pi fL$, and $X_C = 1/2\pi fC$ and at the resonant frequency: $f_r = 1/2\pi \sqrt{LC}$.

SERIES R, L AND C BANDPASS CIRCUITS
X_L and X_C effectively cancel in a series circuit at resonance. For example, in the series circuit of Figure 21-1 the circuit effectively is reduced to $Z_s = R1 + R_{coil}$. Ohms law is used to solve the current I and the ac signal distributed in the circuit. Q(L), which includes the total series resistance, is then used to solve the ac voltages around the circuit, the bandwidth, and both the low and high corner frequencies of the circuit.

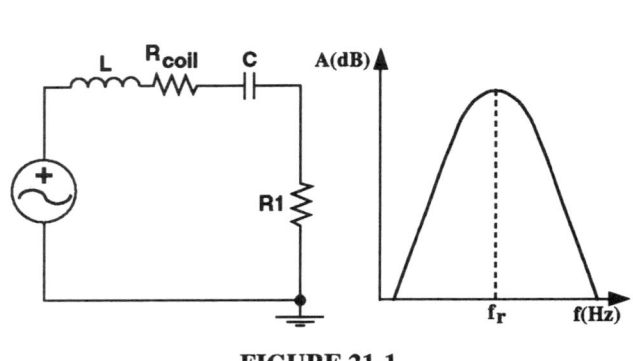

FIGURE 21-1

$$f_r = \frac{1}{2\pi\sqrt{LC}}$$

$$X_L = 2\pi fL$$

$$X_C = \frac{1}{2\pi fC}$$

$$Z_s = R1 + R_{coil}$$

$$I = \frac{V}{R1 + R_{coil}}$$

$$V_{R1} = IR_1$$

$$V_L = IX_L$$

$$V_C = IX_C$$

$$Q(L) = \frac{X_L}{R1 + R_{coil}}$$

$$V_C = Q(L) \times V_{in}$$

$$V_L = Q(L) \times V_{in}$$

$$BW = \frac{f_r}{Q(L)}$$

$$f_L = f_r - \frac{BW}{2}$$

$$f_H = f_r + \frac{BW}{2}$$

$$BW = f_H - f_L$$

PARALLEL LC BANDPASS CIRCUITS
In the parallel LC "tank" circuit of Figure 21-2, the resonant frequency and X_L and X_C solved as shown above for the series RLC circuit. However, the impedance (resistive at resonance) of the parallel LC tank circuit is solved knowing the unloaded Q and the X_L, and the loaded tank impedance must take into account the R1, which is in parallel with the $Z_{p(U)}$. Then Q(L) can be found and the bandwidth and low and high corner frequencies solved.

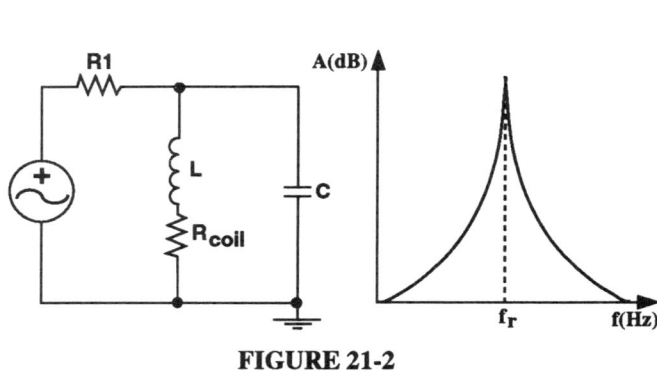

FIGURE 21-2

$$f_r = \frac{1}{2\pi\sqrt{LC}}$$

$$X_L = 2\pi fL$$

$$X_C = \frac{1}{2\pi fC}$$

$$Q(U) = \frac{X_L}{R_{coil}}$$

$$Z_{p(U)} = Q(U) \times X_L$$

$$V_P = \frac{Z_{in} \times Z_{p(U)}}{R1 + Z_{p(U)}}$$

$$Z_{p(L)} = Z_{p(U)} \| R1$$

$$Q(L) = \frac{Z_{p(L)}}{X_L}$$

$$BW = \frac{f_r}{Q(L)}$$

$$f_L = f_r - \frac{BW}{2}$$

$$f_H = f_r + \frac{BW}{2}$$

$$BW = f_H - f_L$$

LABORATORY EXERCISE

READING ASSIGNMENT: Basic Circuit Analysis for Electronics: 309-321.

EXERCISE OBJECTIVES

To become familiar with:

- Series bandpass LC circuits.
- Parallel bandpass LC circuits.
- Series band-reject LC circuits.

PROCEDURE

SECTION I: Series RLC Circuits

PART A: Resonant Conditions (Pre-laboratory Calculations)

1. Analyze the RLC circuit of Figure 21-3(a) at resonant conditions.

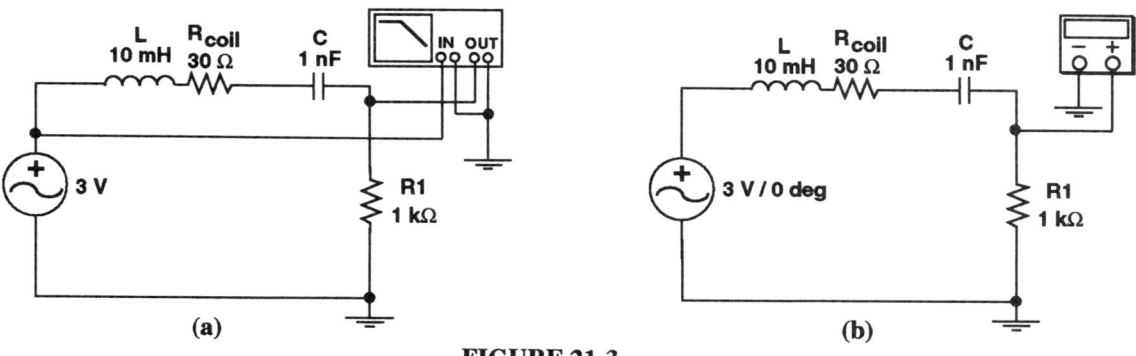

FIGURE 21-3

2. Calculate the resonant frequency (f_r) of the series RLC circuit. Then calculate the voltage drops across R, C, and L.

   a. Calculate the resonant frequency of the series RLC circuit where: $f_r = 1/2\pi\sqrt{LC}$.

   b. Calculate the inductive reactance of L at the resonant frequency where: $X_L = 2\pi fL$.

   c. Calculate the capacitive reactance of C at the resonant frequency where: $X_C = 1/2\pi fC$.

   d. Calculate the approximate series circuit current at the resonant frequency where: $V_{in} = 3$ V and $I = V_{in}/(R1 + R_{coil})$.

   e. Calculate the signal voltage across L at the resonant frequency where: $V_L \approx IX_L$.

   f. Calculate the signal voltage across C at the resonant frequency where: $V_C \approx IX_C$.

   g. Calculate the signal voltage across R at the resonant frequency where: $V_R \approx IR_1$.

   h. Calculate the signal voltage across R_{coil} at the resonant frequency where: $V_{Rcoil} \approx IR_{coil}$.

   i. Calculate the loaded Q of the circuit at the resonant frequency where: $Q(L) = X_L/(R1 + R_{coil})$.

3. Insert the calculated values, as indicated, into Table 21-1.

PART B: Resonant Conditions (Circuit Measurements)

1. Open and select the circuit of Figure 21-3(a) from the file menu, or connect the circuit. The Bode plotter is connected across resistor R1.

a. Set the Bode plotter Initial Frequency (I) to 1 kHz and the Final Frequency (F) at 1 MHz. Then set both the vertical and horizontal to the Log mode, the I at – 40 dB and the F to 0.0 dB.
b. Insert the measured frequency, as indicated, into Table 21-1.

NOTE: The measured resonant frequency is slightly lower than calculated values when using the voltmeter.

2. Open and select the circuit of Figure 21-3(b) from the File menu, or connect the circuit. Set the ac source to 3 V at the measured resonant frequency of 49.7 kHz. At the measured resonant frequency, measure V_{R1}, the voltage across resistor R1.

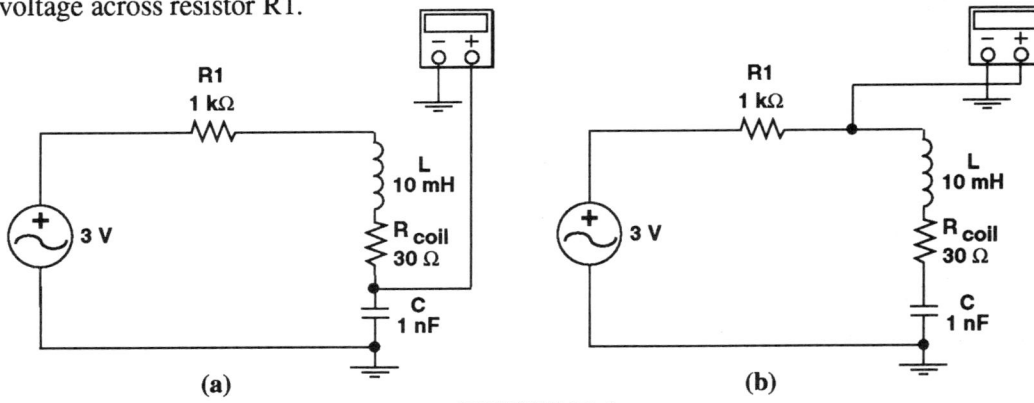

FIGURE 21-4

3. Open and select the circuit of Figure 21-4(a) from the file menu, or connect the circuit. At the resonant frequency, measure V_C, the voltage drop across capacitor C.

4. Open and select the circuit of Figure 21-4(b) from the file menu, or connect the circuit. At the resonant frequency, measure V_{Rcoil}, the voltage drop across the series L and C components.

5. Insert the calculated and measured values, as indicated, into Table 21-1.

TABLE 21-1	f_r	$X_L = X_C$	V_L	Q(L)	V_C	V_R	V_{Rcoil}
CALC.							
MEAS.		/////	/////	/////			

PART B: Bandpass Frequency Response of the Series RLC Circuit

1. Use the Table 21-1 values to solve BW, f_L, f_H, Q(L) and A(dB) of the bandpass circuit of Figure 21-3(a).
   a. Use the measured V_{R1} voltage and the input voltage V_{in} to find the V_{R1}/V_{in} ratio. Then, convert the V_{R1}/V_{in} ratio to db from: $A(dB) = 20 \log V_{R1}/V_{in}$.
   b. Use the V_C and V_{in} values at resonance to calculate the Q or the circuit where: $Q(L) = V_C/V_{in}$.
   c. Use the Q(L) and f_r values to calculate the BW of the circuit where: $BW = f_r/Q(L)$.
   d. Use BW and f_r to find the low corner frequency where: $f_L = f_r - BW/2$.
   e. Use BW and f_r to find the high corner frequency where: $f_H = f_r + BW/2$.

2. Insert the calculated and measured values, as indicated, into Table 21-2.

3. Bode Plotter Setup
   Open and select the circuit of Figure 21-3(a) from the File menu, or connect the circuit. Set the Bode Plotter initial frequency (I) to 1 kHz and the final frequency (F) to 1 MHz. Set both the vertical and horizontal to log mode, the I to – 20 dB, and F to 0.0 dB.

Bandpass and Band Reject LC Filters — 149

a. Activate the circuit and move the cursor to the peak voltage condition, which occurs at the resonant frequency. Measure the frequency and the dB level directly from the Bode plotter.

b. Measure the high corner frequency f_H. Begin at the peak resonant condition and monitor both the frequency and the dB level. Then, increase the frequency by moving the cursor until the amplitude drops to about 3 dB below the peak dB level. Record the approximate f_H frequency.

c. Measure the low corner frequency f_L by moving the cursor to decrease the frequency until the amplitude drops to about 3 dB below the peak dB level. Record the approximate f_L frequency.

d. Use the high and low corner frequencies to verify the bandwidth where: $BW = f_H - f_L$.

e. Use the resonant frequency and bandwidth to solve the Q of the circuit from $Q(L) = f_r/BW$.

4. Insert the calculated and measured values, as indicated, into Table 21-2.

TABLE 21-2	V_{R1}/V_{in}	A(dB)	f_r	f_H	f_L	BW	Q(L)
CALC.			/////				
MEAS.	/////						

SECTION II: Parallel LC Circuits
PART A: Resonant Conditions (Pre-laboratory Calculations)

Analyze the LC tank circuit of Figure 21-5 at resonant conditions. In the circuit a series 100 kΩ resistor buffers the parallel LC tank circuit from the low resistance of the signal generator. Thus, the generator resistance has only minor effect on the Q of the tank circuit. Use a V_{in} of 3 V.

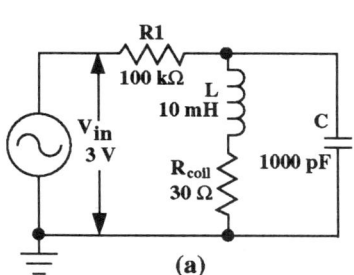
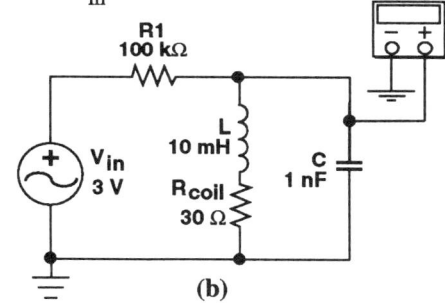

FIGURE 21-5

1. Resonant Frequency Calculations of the Parallel L and C Tank Circuit:

    a. Calculate the resonant frequency of the series L and C circuit where: $f_r = 1/2\pi\sqrt{LC}$.

    b. Calculate the inductive reactance of L at the resonant frequency where: $X_L = 2\pi fL$.

    c. Calculate the capacitive reactance of C at the resonant frequency where: $X_C = 1/2\pi fC$.

    d. Calculate the unloaded Q(U) of the coil where: $Q(U) = X_L/R_{coil}$.

    e. Calculate the approximate tank impedance at the resonant frequency where: $Z_p(U) = Q(U) \times X_L$.

    f. Calculate the voltage across the LC tank circuit at resonance where: $V_{tank} = [V_{in} \times Z_p(U)]/(R1 + Z_p(U))$.

    g. Calculate the V_{tank}/V_{in} ratio. Convert to dB where: $A(dB) = 20 \log V_{tank}/V_{in}$.

2. Circuit Measurements using the Voltmeter: Open and select the circuit of Figure 21-6(b) from the File menu, or connect the circuit.

    a. Activate the circuit and, at the previously measured resonant frequency (49.7 kHz), measure V_{tank}, the voltage drop across the parallel LC tank circuit.

    b. Use the measured V_{tank} voltage and the input voltage V_{in} to the V_{tank}/V_{in} ratio. Then, convert to dB where: $A(dB) = 20 \log (V_{tank}/V_{in})$.

3. Insert the calculated and measured values, as indicated, into Table 21-3.

TABLE 21-3	f_r	$X_L = X_C$	Q(U)	Zp(U)	V_{tank}	V_{tank}/V_{in}	A(dB)
CALCULATED							
MEASURED	/////	/////	/////	/////			

PART B: Bandpass Frequency Response of the Parallel LC Tank Circuit

1. Use the values in Table 21-3 to solve Zp(L), BW, f_L, f_H, and Q(L) of the circuit of Figure 21-6.
   a. Use the Zp(U) and R1 values to calculate the loaded tank impedance where: Zp(L) = Zp(U) || R1.
   b. Use the Zp(L) and X_L values to calculate the loaded Q of the circuit where: Q(L) = Zp(L)/X_L.
   c. Use the Q(L) and f_r values to calculate the BW of the circuit where: BW = f_r/Q(L).
   d. Use BW and f_r to find the low corner frequency where: $f_L = f_r - BW/2$.
   e. Use BW and f_r to find the high corner frequency where: $f_H = f_r + BW/2$.

2. Insert the calculated values, as indicated, into Table 21-4.

3. Bode Plotter Setting and Measurements
   Open and select from the file menu the circuit of Figure 21-6, or connect the circuit. On the Bode plotter set the Initial frequency (I) to 40 kHz and the Final frequency (F) to 60 kHz. Then, set the vertical and horizontal to log, the (I) to – 20 dB, and the (F) to 0.0 dB.
   a. Activate the circuit of 21-6 and monitor the peak response on the Bode plotter, which occurs across the tank circuit at the resonant frequency.
   b. Move the vertical cursor to the peak voltage resonant frequency condition and measure both the frequency and the dB level, which are directly indicated on the Bode plotter.

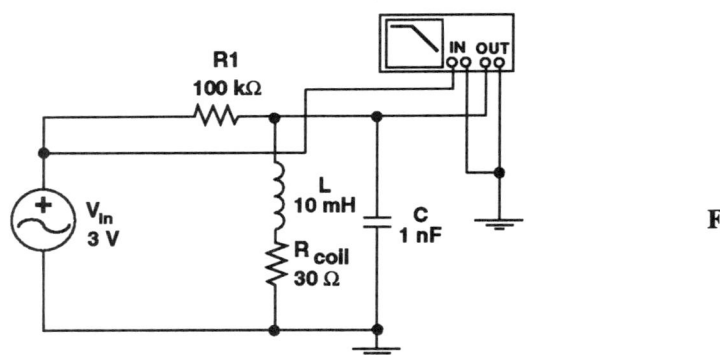

FIGURE 21-6

4. Bandwidth and Circuit Q Measurements
   a. Measure the high corner frequency f_H. Begin at the peak resonant condition and monitor the dB level. Then, increase the frequency until the amplitude drops to about 3 dB below the peak dB level of the resonant condition. Record the f_H frequency.
   b. Measure the low corner frequency f_L by decreasing the frequency until the amplitude drops to about – 3 dB of the peak dB level. Record the f_L frequency.
   c. Use the high and low corner frequencies to solve the bandwidth where: BW = $f_H - f_L$.
   d. Use the resonant frequency and bandwidth to solve the circuit Q from: Q(L) = f_r/BW.

5. Insert the calculated and measured values, as indicated, into Table 21-4.

TABLE 21-4	Zp(L)	A(dB)	f_r	f_H	f_L	BW	Q(L)
CALCULATED		/////	/////				
MEASURED	/////						

Bandpass and Band Reject LC Filters — 151

PART C: Plotting the Bandpass Response of the Parallel LC Tank Circuit

Use the response on the Bode plotter to obtain f_r, f_L, f_H, and the low and high frequencies that occur at the −20 dB levels. Measure the frequency directly from the Bode plotter. The cursor can be dragged or incrementally moved by using the arrows located on the Bode plotter. So move the cursor as close as possible to each of the respective calculated frequencies and find both the measured frequency and the associated dB level.

1. To provide a reference, insert the calculated f_r, f_L, and f_H frequencies into Table 21-5.

2. To measure f_r, slide the cursor to the peak voltage condition. Record by f_r and the A(dB). Read directly from the Bode plotter.

3. To measure f_L and f_H, slide the cursor to the A(dB) level, about 3 dB lower than the peak voltage condition.

4. At −20 dB measure both the low and high frequencies.

5. Insert the measured frequency and the output voltage values in dBs, as indicated, into Table 21-5.

TABLE 21-5	f(low) at −20 dB	f_L	f_r	f_H	f(hi) at −20 dB
CALC. FEQUENCY	/////				/////
MEAS. FREQUENCY					
MEASURED A(dB)	−20 db				−20 db

6. Plotting the Frequency Response
   Use the graph of Figure 21-7 and the A(dB) data from Table 21-5 at each of the selected frequencies. Plot a frequency response of A(db) versus frequency, where the response should "roll on", peak, and "roll off" as the frequency is increased from below to above the resonant frequency.

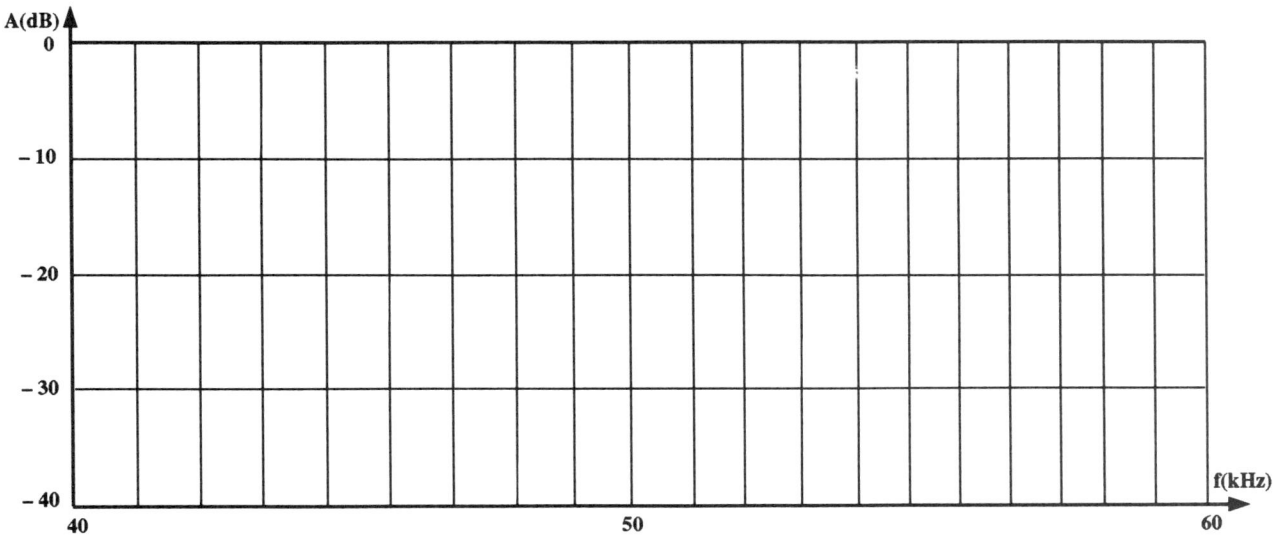

FIGURE 21-7

SECTION III: Band-Reject Frequency Response of the Series RLC Circuit

1. Use the values in Table 21-1 to solve BW, f_L, f_H, Q(U) and A(dB) of the band reject circuit of Figure 21-8.
   a. Use the measured V_{Rcoil} voltage and the input voltage V_{in} at the previously calculated resonant frequency

to find the V_{Rcoil}/V_{in} ratio. Then, convert to dB where $A(dB) = 20 \log (V_{Rcoil}/V_{in})$.

b. Use the V_C and V_{Rcoil} values at resonance to calculate the Q of the coil where: $Q(U) = V_C/V_{Rcoil}$.

c. Use the $Q(U)$ and f_r values to calculate the BW of the coil where: $BW = f_r/Q(U)$.

d. Use BW and f_r to find the low corner frequency where: $f_L = f_r - BW/2$.

e. Use BW and f_r to find the high corner frequency where: $f_H = f_r + BW/2$.

f. Insert the calculated values, as indicated, into Table 21-6.

2. Bode Plotter Setup

Open and select the circuit of Figure 21-8 from the File menu, or connect the circuit. Use the Bode plotter to provide the frequency response. Set the Bode plotter Initial frequency (I) to 40 kHz and the Final frequency (F) to 60 kHz. Set the vertical and horizontal to log mode, the I to – 40 dB, and F to 0.0 dB.

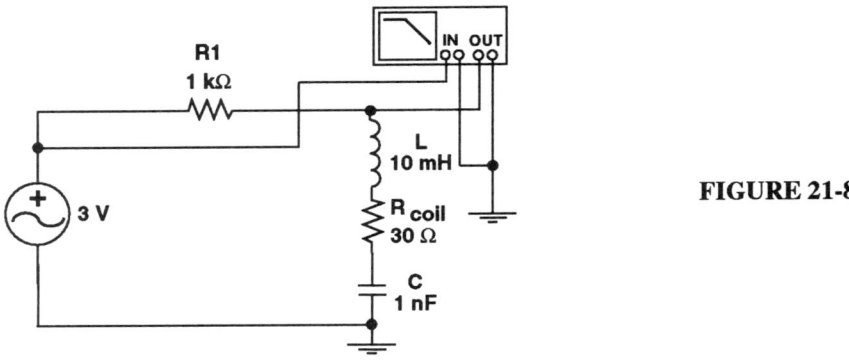

FIGURE 21-8

a. Activate the circuit and move the cursor to the "dip" voltage condition, which occurs at the resonant frequency. Measure the frequency and the dB level directly on the Bode plotter.

b. Measure the high corner frequency f_H. Begin at the dip resonant condition and monitor the dB level. Then, increase the frequency until the amplitude rises to about 3 dB above the dip dB level. Record the approximate f_H frequency.

c. Measure the low corner frequency f_L by decreasing the frequency until the amplitude rises to about 3 dB above the dip dB level. Record the approximate f_L frequency.

d. Use the high and low corner frequencies to solve the bandwidth where: $BW = f_H - f_L$.

e. Use the resonant frequency and bandwidth to solve the Q of the coil from $Q(U) = f_r/BW$.

3. Insert the calculated and measured values, as indicated, into Table 21-6.

TABLE 21-6	V_{Rcoil}/V_{in}	A(dB)	f_r	f_H	f_L	BW	Q(U)
CALC.							
MEAS.	/////						

Questions and Problems: Basic Circuit Analysis for Electronics: 326-327

CHAPTER 22

COMPLEX WAVESHAPES

INTRODUCTION

In electronics, it is the information contained in the waveshapes, both symmetrical and non-symmetrical, sine wave and non-sine wave, that requires study. A voltmeter is generally used to measure the rms value of a signal voltage, while the oscilloscope can provide visual measurement of the peak-to-peak voltage of the waveshape. The formulas for converting peak-to-peak voltages to rms and dc voltages of the symmetrical sine wave, square wave, and triangle wave are shown in conjunction with the waveshapes in Figure 22-1.

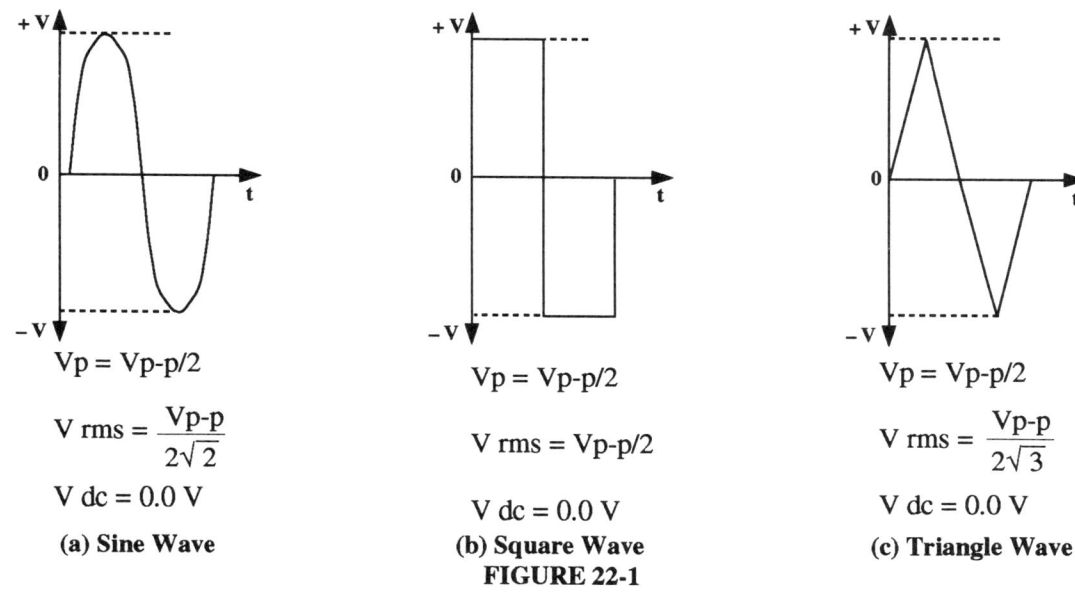

(a) Sine Wave (b) Square Wave (c) Triangle Wave

FIGURE 22-1

HALF-WAVE FORMULAS (SINE WAVE AND SQUARE WAVE)

A symmetrical sine or square wave chopped in half, where only the positive-going wave exists, provides a positive dc voltage that is the average of the positive-going wave with reference to 0.0 V. The V_p, V rms, and V dc formulas for the one-half sine wave are shown in Figure 22-2(a) and those for the one-half square wave are shown in Figure 22-2(b).

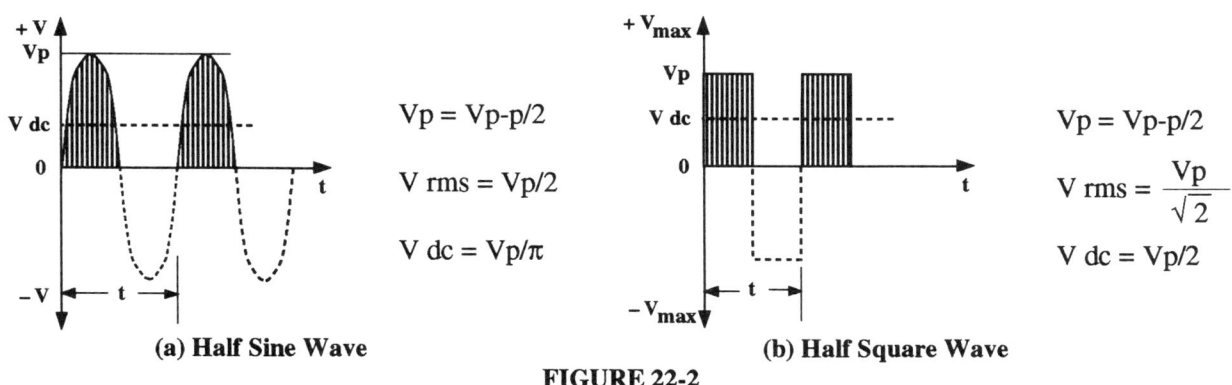

(a) Half Sine Wave (b) Half Square Wave

FIGURE 22-2

153

LABORATORY EXERCISE

READING ASSIGNMENT: Basic Circuit Analysis for Electronics: 328-336.

EXERCISE OBJECTIVES

To become familiar with:

- Symmetrical waveshapes (sine, square, and triangle).
- Non-symmetrical waveshapes (sine and square).

PROCEDURE

SECTION I: Comparing Symmetrical Sine, Square, and Triangle Waves

PART A: The Sine Wave

1. In the connection of Figure 22-3 the function generator is switched to the sine wave functions and set at 12 Vp-p and 1 kHz.
   a. Calculate the rms voltage of the 12 Vp-p sine wave where: V rms = Vp-p/$2\sqrt{2}$.
   b. Calculate the dc voltage of the symmetrical sine wave.

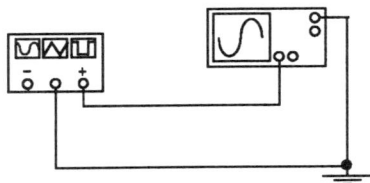 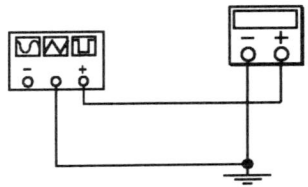

(a) Function Generator and Oscilloscope   (b) Function Generator and Multimeter

FIGURE 22-3

2. Open and select the circuit of Figure 22-3(a) from the File menu, or connect the circuit.
   a. Set the amplitude of the sine wave on the oscilloscope to 12 Vp-p, where the volts/div settings are 2 V/div and the time/div is set at 0.5 ms/div. Set the generator to 1 kHz, 6 Vpeak, and sine wave.
   b. Measure the positive-going peak voltage of the sine wave on the oscilloscope.

3. Open and select the circuit of Figure 22-3(b) from the File menu, or connect the circuit.
   a. Set the voltmeter to volts (V) and AC (~) mode and measure the rms voltage on the voltmeter.
   b. Set the voltmeter to volts (V) and DC (—) mode and measure the dc voltage, if any, on the voltmeter.

NOTE: The generator is set to 6 volts peak voltage and sine wave to obtain a 12 Vp-p sine wave on the scope.

4. Insert the calculated and measured values, as indicated, into Table 22-1.

PART B: The Square Wave

1. In the connection of Figure 22-3(a), the function generator is switched to the square wave function and set at 12 Vp-p (6 Vpeak) and 1 kHz.
   a. Calculate the rms voltage of the 12 Vp-p wave where: V rms = Vp-p/$\sqrt{2}$.
   b. Calculate the dc voltage of the symmetrical square wave.

2. Set the amplitude of the square wave on the oscilloscope to 12 Vp-p, where the volts/div settings are set to 2 V/div and the time/div at 0.5 ms/div. Set the generator to 1 kHz, 6 Vpeak, and sine wave. Measure the positive-going peak voltage of the square wave on the oscilloscope.

3. Open and select the circuit of Figure 22-3(b) from the File menu, or connect the circuit.
   a. Set the voltmeter to volts (V) and AC (~) mode and measure the rms voltage on the voltmeter.
   b. Set the voltmeter to volts (V) and DC (—) mode and measure the dc voltage, if any, on the voltmeter.

4. Insert the calculated and measured values, as indicated, into Table 22-1.

PART C: The Triangle Wave

1. In the connection of Figure 22-3(a), the function generator is switched to the triangle wave function and set at 12 Vp-p (6 Vpeak) and 1 kHz.
   a. Calculate the rms voltage of the 12 Vp-p triangle wave where: V rms = Vp-p/2 $\sqrt{3}$.
   b. Calculate dc voltage of the symmetrical triangle wave.

2. Set the amplitude of the triangle wave on the oscilloscope to 12 Vp-p, where the volts/div settings are set to 2 V/div and the time/div at 0.5 ms/div. Set the generator to 1 kHz, 6 Vpeak, and triangle wave. Measure the positive-going peak voltage of the triangle wave on the oscilloscope.

3. Open and select the circuit of Figure 22-3(b) from the File menu, or connect the circuit.
   a. Set the voltmeter to volts (V) and AC (~) mode and measure the rms voltage on the voltmeter.
   b. Set the voltmeter to volts (V) and DC (—) mode and measure the dc voltage, if any, on the voltmeter.

4. Insert the calculated and measured values, as indicated, into Table 22-1.

TABLE 22-1	Sine Wave			Square Wave			Triangle Wave		
	Vp	V rms	V dc	Vp	V rms	V dc	Vp	V rms	V DC
CALCULATED									
MEASURED									

SECTION II: Comparing Non-symmetrical Waveshapes
PART A: The Half-Wave Sine Wave

1. In the half-wave circuit connection of Figure 22-4, the function generator is switched to the sine wave function and set to 12 Vp-p and 1 kHz.

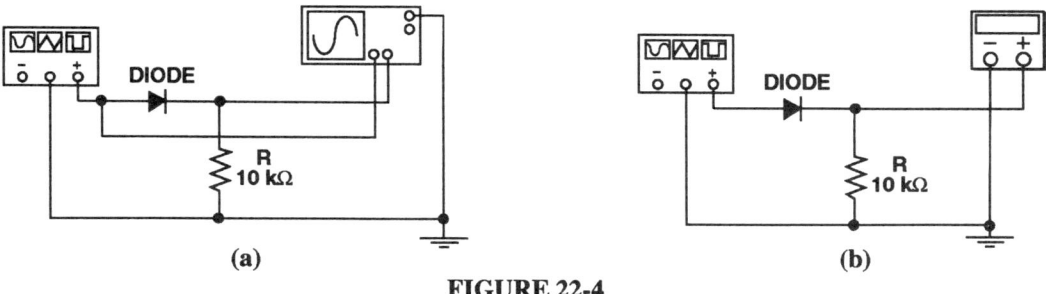

FIGURE 22-4

2. Calculate the voltages at the load.
   a. Calculate the peak voltage at the load where: Vp = Vp-p/2 – V_F.
   b. Calculate the dc voltage at the load where: V dc = Vp/π ≈ 0.318 Vp.
   c. Calculate the ac voltage at the load where: V ac ≈ 0.386 Vp.
   d. Calculate the rms voltage at the load where: V rms = Vp/2.

NOTE: The rms voltage can also be solved from V rms = $\sqrt{V dc^2 + V ac^2}$ formula. Also, V_F ≈ 0.7 V.

3. Measurements using the Oscilloscope:
   a. Open and select the circuit of Figure 22-4(a) from the File menu, or connect the circuit. Use the oscilloscope and set the input sinewave amplitude to 12 Vp-p. Typical volts/div settings are at 2 Vp-p/div and time/div is set at 0.5 ms/div.
   b. Measure both the peak-to-peak voltage of the generator and the positive-going peak voltage (wave) across the load resistor.

4. Measurements using the Voltmeter:
   a. Select from the File menu the circuit of Figure 22-4(b) and open, or connect the circuit.
   b. Set the voltmeter to DC mode and measure the dc voltage across the load.
   c. Set the voltmeter to the AC mode and measure the ac voltage across the load.
   d. Use the measured V dc and V ac values to find V rms at the load from $V\,rms = \sqrt{V\,dc^2 + V\,ac^2}$.

NOTE: The peak voltage at the load effectively chops the input peak-to-peak voltage in half, minus the 0.7 V of the forward biased diode. So, the slightly less than half wave at the load will have both an ac and a dc voltage content that combine to provide the rms voltage.

PART B: The Square Wave

1. In the half-wave connection of Figure 22-4(a), the function generator is switched to the square-wave function and set to 12 Vp-p and 1 kHz.
   a. Calculate the maximum (peak) voltage of the positive-going wave at the load where: $Vp = Vp\text{-}p/2 - V_F$.
   b. Calculate the dc voltage of the maximum (peak) wave at the load where: $V\,dc = Vp/2$.
   c. Calculate the ac voltage of the maximum (peak) wave at the load where: $V\,ac = Vp/2$.
   d. Calculate the rms voltage at the load where: $V\,rms = Vp/\sqrt{2}$.

2. Measurements using the oscilloscope:
   a. Set the input sine wave amplitude to 12 Vp-p, with the volts/div set on 2 V/div and the time/div set on 0.5 ms/div.
   b. Measure both the peak-to-peak voltage of the generator and the positive-going peak voltage (wave) across the load resistor.

3. Measurements using the voltmeter:
   a. Set the voltmeter to the DC mode and measure the dc voltage across the load.
   b. Set the voltmeter to the AC mode and measure the ac voltage across the load.
   c. Use the measured V dc and V ac values to find V rms at the load from $V\,rms = \sqrt{V\,dc^2 + V\,ac^2}$.

4. Insert the calculated and measured values into Table 22-2.

TABLE 22-2	Halfwave Sine Wave					Halfwave Square Wave				
	Vp-p	Vp	V dc	V ac	V rms	Vp-p	Vp	V dc	V ac	V rms
CALC.										
MEAS.										

Questions and Problems: Basic Circuit Analysis for Electronics: 340

CHAPTER 23
SQUARE WAVE TESTING OF THE CORNER FREQUENCIES

INTRODUCTION
In measuring the corner frequency of low or high pass filter circuits either sweep measurements or square wave testing can be used. The advantages of square wave testing are a fixed frequency is used, it is quick and accurate, and the procedure is basically the same regardless of how simple or complex the circuit.

HIGH CORNER FREQUENCY
In the low pass filter circuit, using swept measurements, the amplitude of the signal across the output rolls off and the amplitude drops to 0.707 of V_{in} at the high corner frequency. So, the measured high corner frequency occurs at 0.707 of V_{in} and is calculated from $f_H = 1/2\pi RC$. For the circuit of Figure 23-1(a), using square wave testing as shown in Figure 23-1(b), the rise time is measured and the high corner frequency found from $f_H = 0.35/t_r$.

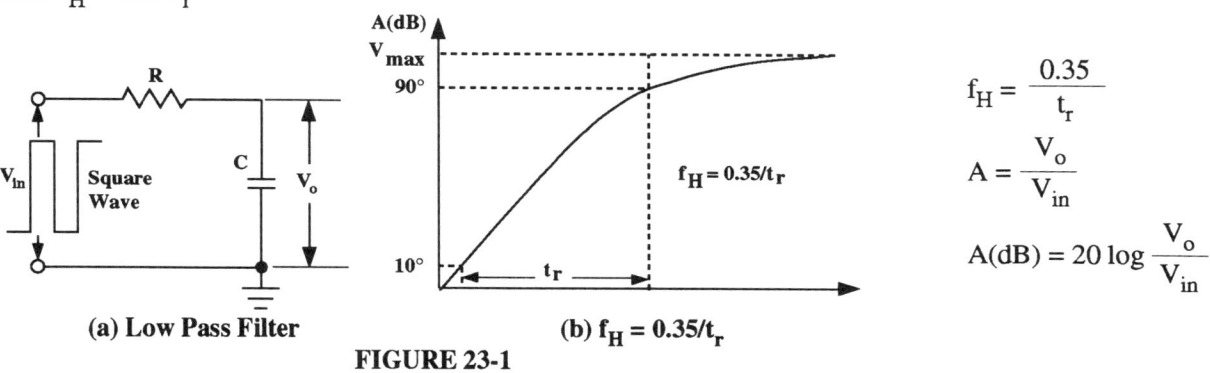

FIGURE 23-1

LOW CORNER FREQUENCY
In the high pass filter circuit of Figure 23-2(a), using swept measurements, the signal across the output rolls on and the amplitude increases to 0.707 of V_{in} at the high corner frequency. Then, the high corner frequency is calculated from $f_H = 1/2\pi RC$. In square wave testing a square wave applied to the RC high pass filter circuit provides a response similar to that shown in Figure 23-2(b). Tilt is found by measuring V1 and V2 and calculating from: tilt = (V1 – V2)/V1. Once the tilt is measured, the low corner frequency is found from: f_L = tilt × f_o/π.

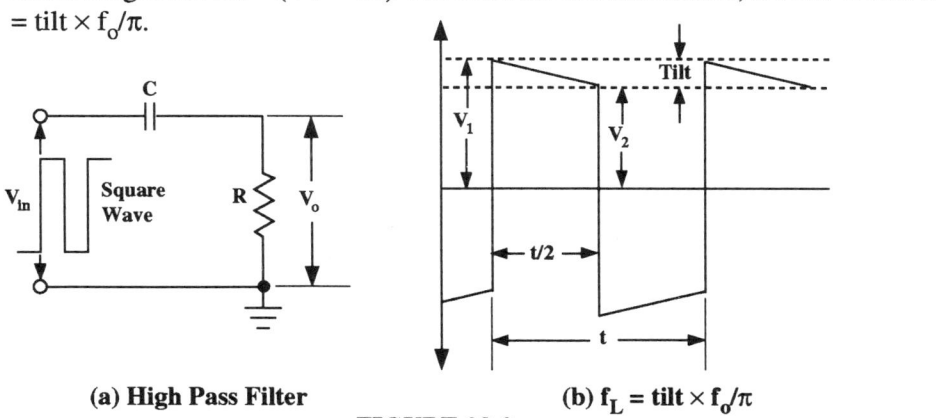

FIGURE 23-2

LABORATORY EXERCISE

READING ASSIGNMENT: Basic Circuit Analysis for Electronics: 341-346.

EXERCISE OBJECTIVES

To become familiar with:

- Measuring rise time and solving f_H using square wave testing.
- Measuring tilt and solving f_L using square wave testing.

PROCEDURE

SECTION I: High Corner Frequency Analysis—Square Wave Testing

1. Analyze the low pass RC circuit of Figure 23-3.

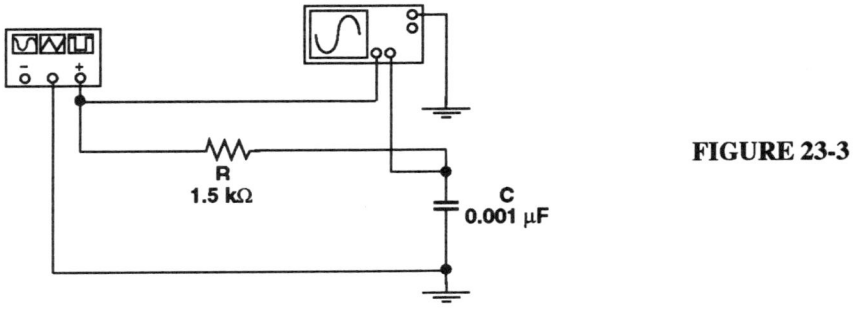

FIGURE 23-3

2. Calculate the high corner frequency, where $f_H = 1/2\pi RC$, $R = 1.5$ kΩ, and $C = 0.001$ μF.

3. Measure rise time (t_r) and derive f_H. Open and select the circuit of Figure 23-3 from the File menu, or connect the circuit. Set the scope to 1 V/div and 10 μs/div. Use an input square wave voltage of 5 Vp-p at approximately 10 kHz and DC mode.
    a. Monitor the output voltage across the output capacitor, and measure the rise time t_r which occurs between the 10% and 90% points.
    b. Use the measured rise time to calculate the high corner frequency, where $f_H = 0.35/t_r$.

NOTE: The generator is set to 2.5 V to obtain the 5 Vp-p input square wave on the scope. For 5 Vp-p the rise time should occur between the -2 V (10%) and 2 V (90%) points.

4. Insert the calculated and measured values, as indicated, into Table 23-1.

TABLE 23-1	f_H	V_{in}	t_r	f_H
CALCULATED				
MEASURED				

5. Sketch the square wave input into the graph of Figure 23-4(a) and the rise time into the graph of 23-4(b). Label properly. Indicate the 10% and 90% points on the rise time curve. Also, indicate the scope settings.

Square Wave Testing of the Corner Frequencies —159

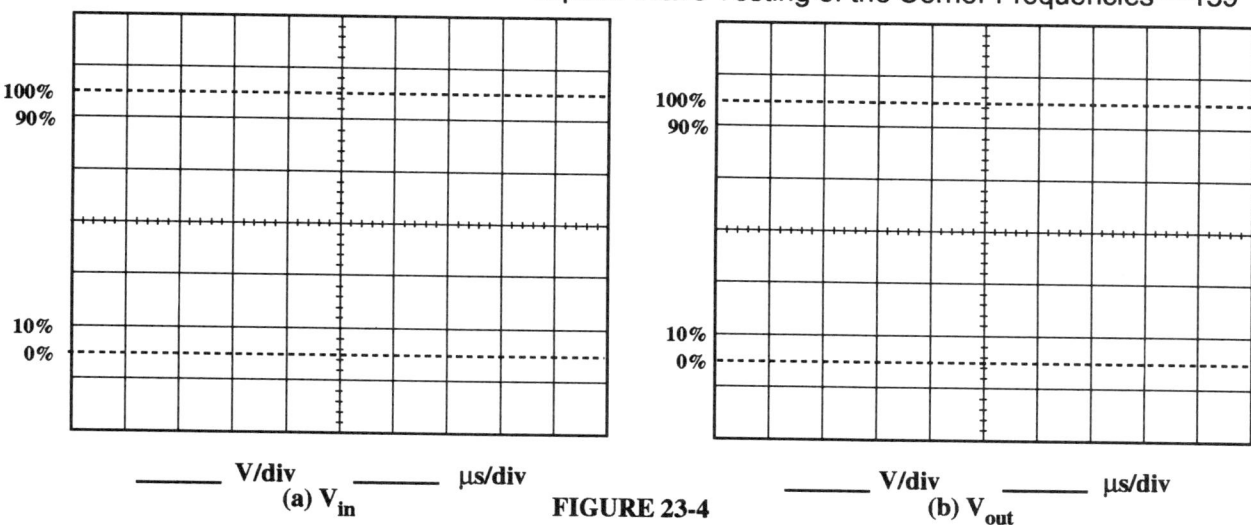

_____ V/div _____ μs/div
(a) V_{in}

FIGURE 23-4

_____ V/div _____ μs/div
(b) V_{out}

SECTION II: Low Corner Frequency Analysis—Square Wave Testing

1. Analyze the low pass RC circuit of Figure 23-5.

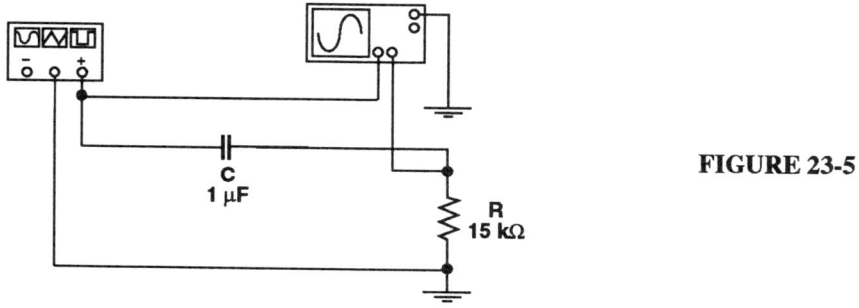

FIGURE 23-5

2. Calculate the corner and operating frequencies:

   a. Calculate the corner frequency where: $f_L = 1/2\pi RC$, R = 15 kΩ, and C = 1 μF.

   b. Calculate operating frequency at the approximate tilt of 10% where: $f_o = 10\pi f_L$.

3. Measure Tilt and derive f_L. Open and select the circuit of Figure 23-5 from the File menu, or connect the circuit. Set the scope to 1 V/div, 0.5 ms/div, and to DC mode. Use an input square wave voltage of 6 Vp-p at the calculated operating frequency and center the wave input on the oscilloscope.

   a. Monitor the output voltage across the output resistor and measure V_1 and V_2, measured with respect to the 0.0 V reference. Use the vertical cursors to measure V1 and V2.

   b. Calculate the tilt where: tilt = (V1 – V2)/V1.

   c. Use the measured tilt and the generator operating frequency (f_o) to calculate the low corner frequency, where: $f_L = \text{tilt} \times f_o/\pi$.

4. Insert the calculated and measured values, as indicated, into Table 23-2.

TABLE 23-2	f_L	f_o	V_{in}	V_1	V_2	tilt	f_L
CALCULATED							
MEASURED							

160 — Basic Circuit Analysis For Electronics Using Electronic Workbench®

5. Sketch the square wave input wave on the graph of Figure 23-6(a) and output wave on the graph of Figure 23-6(b). Label properly. Indicate the V_1 and V_2 amplitudes and the actual V/div and ms/div scope settings.

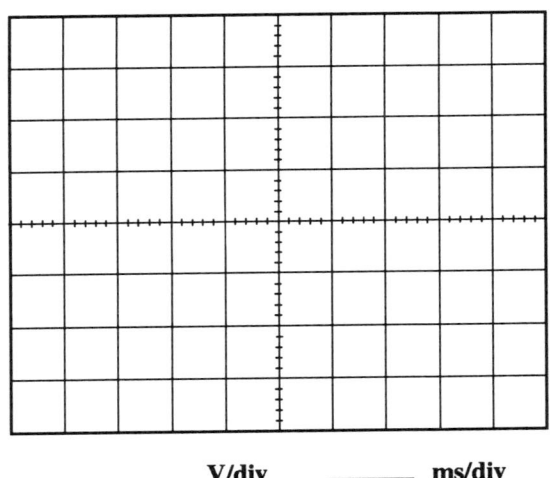
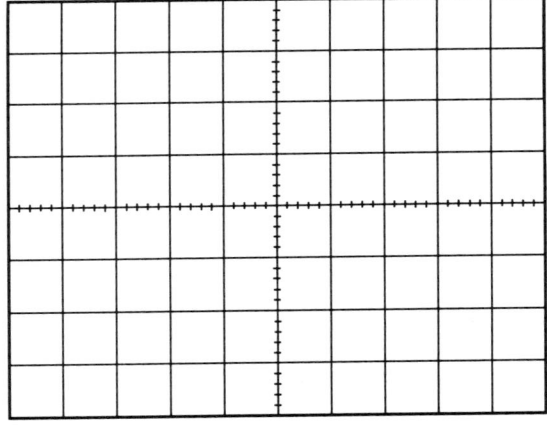

_____ V/div _____ ms/div
(a) V_{in}

_____ V/div _____ ms/div
(b) V_o

FIGURE 23-6

Questions and Problems: Basic Circuit Analysis for Electronics: 349-350